A Test of Morals

Surgical, Ethical, and Psychosocial Considerations in Human Head Transplantation

L. Allen Furr
Professor Emeritus
Auburn University

CRC Press
Taylor & Francis Group
Boca Raton London New York

CRC Press is an imprint of the
Taylor & Francis Group, an **informa** business

A SCIENCE PUBLISHERS BOOK

First edition published 2024
by CRC Press
2385 NW Executive Center Drive, Suite 320, Boca Raton FL 33431

and by CRC Press
4 Park Square, Milton Park, Abingdon, Oxon, OX14 4RN

CRC Press is an imprint of Taylor & Francis Group, LLC

Library of Congress Cataloging-in-Publication Data (applied for)

ISBN: 978-1-032-43366-0 (hbk)
ISBN: 978-1-032-43371-4 (pbk)
ISBN: 978-1-003-36701-7 (ebk)

DOI: 10.1201/9781003367017

Typeset in Palatino Linotype
by Radiant Productions

Dedication

To my sister Saidie Ree who really is
my elder despite what she says.

Preface

One day in 2016, John Barker, one of the early pioneers of modern composite tissue allotransplantation (CTA) and with whom I, along with his team of clinicians and researchers in the early 2000s, had collaborated on a number of papers on various ethical and psychosocial aspects of face transplantation, contacted me with an interesting proposal. That John was calling with a novel idea was nothing new as his spirit of inquiry is best described as a dynamic state of perpetual intellectual motion. Oftentimes his imaginings are beyond my reach (but brainstorming with John never fails the promise of a good time and occasionally involves having a go at new endeavors, such as our aborted attempt at documentary filmmaking), but this one was attractive, albeit in a somewhat macabre fashion. After the exchange of a few niceties, he asked if I had heard of head transplantation, and, of course, I said, no. Afterall, who had? He explained the concept to me and said something to the effect of "saddle up, we're going to write a couple of papers on it." As is always the case with John, when he identifies a target for his voracious curiosity, things happen. He assembled a team of expert scholars, though I did not consider myself as one of them, and divided the labor along lines of expertise, and off we went.

John recruited Mark Hardy, an endowed professor of surgery and internationally renowned transplantation specialist at the New York-Presbyterian Hospital, and Juan Barrett, a prominent Spanish plastic surgeon who had performed multiple face transplant operations and was widely published in medical journals. Together we wrote three papers that were published in 2017 and 2018, which collectively have been cited over 40 times, as of summer 2023 and per Google Scholar. These papers were among the first comprehensive academic reviews of head transplantation, focusing not only on ethics but on immunology and surgical procedures as well.

After publishing these papers, we dropped the subject and returned to our regular research agendas. Others, as we soon learned, were also on the trail of head transplantation. As our papers were coming out, the *American Journal of Bioethics Neuroscience* released a special edition on head transplantation in 2017 as did the *Journal of Medicine and Philosophy* a few years later in 2022. The latter publication extended the ethical debate to include the experiment's implications on a wide range of topics such as cross-cultural definitions of the self, the meaning and desirability of immortality, how we, the embodiment of a society, define death, and whether head transplantation is even legal. From 2017 onward, other articles on the ethics and techniques of transplanting a human head began to appear in journals in different disciplines. Dedicating the insights of this many scholars to this subject suggests head transplantation has triggered the philosophical and medical imagination in ways beyond age-old thought experiments and hypothetical scenarios in which one person's brain was placed into another's body.

Scholars and clinicians were not alone in thinking about head transplantation. Since Mary Shelley, as some contend, invented modern science fiction with her 1818 publication of *Frankenstein* (while "Frankenstein" is a term often employed somewhat incorrectly to describe and criticize head transplantation, Shelley was really writing about the reanimation of dead tissues and assembling them into a new "creature"), writers have relied on ghoulish plot devices like brain transfers or transplanting heads. One of the first was *Professor Dowell's Head*, written in 1925 by Russian author Alexander Belyaev, in which a young girl develops a relationship with a severed head kept alive artificially. Following Belyaev's lead, the mid-20th Century witnessed a rash of brain transplant stories. The original *Star Trek* television show featured an episode in 1968 in which aliens remove Spock's brain from his body and plug it into the computer infrastructure that controls their civilization. The plot of the 1972 French film *Man with the Transplanted Brain* (*L'Homme au cerveau greffé*), centers around a neurosurgeon's attempt to avoid death from heart disease by transplanting his brain into the head of younger, healthier man. On a similar theme, Robert Heinlein's novel, *I Will Fear No Evil*, tells the story of a rich but aging man who pays a large sum to have his brain transplanted to the body of a younger person. The body donor, however, was a woman whose memories and personality were carried by her body into the brain of the man. Another is the lesser known 1973 novel *Friends Come in Boxes* by Michael Coney in which the population explosion, a much feared social problem when the book was written, was solved by transferring an adult's brain into a baby's body upon reaching 40 years of age, thus achieving zero net population growth. Negative growth followed because some people would die of natural causes or accidents before reaching 40, and the governing regime would not allow people to "transfer" if they broke the law. Other films, such as Mel Brooks' *Young Frankenstein* in 1974 and Jordan Peele's *Get Out* later in 2017, and stories such as Frederik Pohl's "The Tunnel Under the World" published in 1955, among many others, utilized variations on the brain transplant theme to work out their narratives.

But art imitating life imitating art notwithstanding, head transplantation had already been explored "for real" by the time many of these stories were concocted. Charles Guthrie in 1908 transplanted the head of a dog onto the neck of a second dog, and in 1954, Soviet transplantation innovator Vladimir Demikhov announced that he had successfully attached the head of one dog onto the body of another, creating a functioning albeit short-lived two-headed canine. American Robert White as late as the 1970s conducted a series of operations in which he fully transplanted a monkey's head onto another's body. The head could see, hear, taste, and smell, though the monkey was paralyzed from the neck down. The works of these three experimenters achieved relatively minor newsworthiness and were not widely known among the public or even physicians and researchers. Nonetheless, as we will see in Chapter 2, their work did not progress in a social vacuum. Many scholars were aware of their experiments, and several were quick to pass judgement on the moral architecture of transplanting a human head. Still, head and brain transplantation were vastly more popular in the realm of fiction while the genuine attempts to transplant a head were largely hidden from the public's gaze, or at least never entered the collective consciousness.

What accounts for physicians' and academicians' sudden interest in head transplantation after having largely ignored the idea for the three decades following White's animal experiments? Answer: in 2013 Italian neurologist Sergio Canavero declared that he had concocted a surgical protocol that he claimed would enable the success of an attempt to transplant a living head onto another's body. He also announced that he would perform the first such transplant by December of 2017 on a terminally ill Russian individual who had volunteered to undergo that first attempt. Canavero, inspired by the work of Robert White, promised that his innovations would allow this patient to stave off death by removing his head from his dying body and re-attaching it to a brain-dead donor's body whose head would be similarly surgically removed. The announcement was covered by news agencies throughout the world, and Canavero was interviewed numerous times in several countries. These interviews notwithstanding, the public would remain largely "out of the loop" and would not immediately hear of his impending experimental surgery. Research scholars in medicine, clinicians, and bioethicists, especially in philosophy, however, took notice and began the process of critically evaluating both the techniques and the ethical implications of Canavero's plan.

It is my purpose in writing this book to pull these assessments, both for and against, of Canavero's proposal into one location and to strive for a conclusion on the moral legitimacy of human head transplantation. As we will see, transplanting a head is different from all other transplants in terms of its feasibility, potential for negative psychological outcomes, and conflict with established social norms and values relating to conventional ethics and embodied identity. Legal complications are also in play, and arguments that state that informed consent is unattainable within the current state of knowledge appear insurmountable, which implies an immediate ethical roadblock to progressing with the experiment.

Despite the criticisms, Canavero and his colleagues have persisted in moving forward. They continue to conduct research and talk to the media when the opportunity arises, though that first volunteer withdrew his offer to undergo the surgery and 2017 passed without any indication that the surgery happened. It is not clear if the volunteer's withdrawal was the reason that the operation did not occur, and why Canavero has yet to announce a new target date is similarly uncertain. Nonetheless, Canavero's theory and promotion of head transplantation have intrigued others, and a small body of scientific research has emerged that provides suggestive evidence that his proposal might work.

Despite this postponement, some writers have said that a head transplant will happen eventually, either by Canavero or someone else, and perhaps not for another 50 years or longer. Therefore, this book is not necessarily about Canavero, though it is near impossible to separate his name from the concept of head transplantation: his identity and his work are now indistinguishable. It is about the ethics of transplanting a human head and, for that matter, animal heads and how such a theory is promoted and debated. Because Canavero's theory is the only known procedure for this transplant that is known in the scholarly literature, I am obligated to follow his vision, and that of his team, and his presence in the world as a social actor who is attempting a paradigm shift in not only transplant medicine but in our thinking about life and death, the meaning of the human body, and our understanding of the relationship between our bodies and social identity. Head transplantation is more than a medical novelty. It has implications for our grasp of the links between neurology, psychology, and sociology, what is often called the biopsychosocial self—the interconnectedness of mind, body, and socio-cultural context. If it were to be successful in the way Canavero predicts, transplanting heads will challenge current biopsychosocial knowledge. If not, it may reinforce current beliefs that suggest that the neurological disruptions involved in head transplantation will have a profoundly negative psychosocial consequence for recipient patients. Will receiving a new body have essentially no impact on the self, which is Canavero's expectation, or will it result in madness, as others have predicted? This book will strive to integrate these ideas and place them in the field of established neurology, psychology, and sociology to reach a conclusion on the ethical value of Canavero's proposal and allow readers to settle the matter in their own minds.

Chapter 1 will start at the beginning and introduce head transplantation in the general sense, that is, to readers who are unfamiliar with the concept. Transplanting heads remains relatively unknown and continues to surprise people when they first hear that researchers are exploring the possibility. In this chapter, we will review the theoretical orientation used to make sense of the myriad of ethical concerns raised by the experiment. Utilizing a practical approach implies an inductive reasoning akin to grounded theory in which we begin with observations of the phenomenon under study and then formulate evidence-driven conclusions about the ethical value of that phenomenon. The goal of the book, therefore, is to derive a moral position on head transplantation based on the current state of knowledge and to fill in the gaps, where possible, where present information is lacking.

Chapter 2 reviews the possible benefits and risks of head transplantation. The benefits list is relatively short: it will, theoretically, prevent a mortally ill person from dying. That outcome, however, is not free from peril. A patient may survive, but there are risk considerations that accompany survival, and these risks are not trivial. Many have expressed doubts about the potential fitness of a survivor. Countless things can go wrong in a surgery of this magnitude. Virtually every step of the proposed protocol is an independent surgery outside of the transplantation theater, and there are many such steps. An error or miscalculation, or the discovery of failure of concept at any juncture of the surgery can trigger catastrophic results, even for patients who survive. There are many decisions to be made in removing and maintaining a living head and reattaching it to a body that will be surgically decapitated and prepped to receive a new head. And that is just the surgical part of the process. The social and psychological implications of this proposal are unexplored and untestable prior to the first surgery. There is no way to predict what will happen to the new person's psychological well-being, social identity, and mind-body agreement. Any informed consent offered to the first volunteers must state that future biopsychosocial dynamics are unidentified and perhaps unidentifiable. One of the objectives of this chapter is to explore what awaits the transplanted person's psychological and sociological constitution and functionality.

Chapter 3 targets the ethics of head transplantation. Each of the risks and outcomes presented in Chapter 2 poses its own set of ethical questions. Consequently, because the risk list is long, so is the inventory of yet unresolved ethical quandaries, many of which seem unresolvable. A book on the ethics of head transplantation could easily be 150 pages of nothing but questions. It seems that many academic articles devote as much text to what we need to resolve as they do towards the resolution of those queries. Ethical challenges to head transplantation range from its legality (some say it equates to murder) to the likelihood that the resulting person will be insane. The objective here is to pull all the problems together and search for an answer to them or, at least, to assemble the current opinions and report on their salience.

The concluding Chapter 4 connects culture and the body and draws conclusions about the ethical merits of transplanting heads. How will society's reaction influence the future of this medical novelty? What does this mean about the intersection of the meaning of the body, culture, and technology? Does the ethical profile of head transplantation mesh with social definitions of the body and the role of medicine in post-modern culture? Here we will try to reach a conclusion on the social definitions of this particular reality and see which way the moral compass is pointing.

Sociology is a new player in the debate on head transplantation largely, in part, because the discipline has historically not been involved in the grander world of bioethics. That is beginning to change, and head transplantation offers an opportunity for sociologists to play a major role in facilitating a collective decision on its moral value. Throughout the pages to follow, sociological knowledge will help us understand several aspects of head transplantation. I will rely on the sociology of body literature to understand how individuals intersect with society at large in a physical or material way. In addition, a sociological approach will inform our understanding of identity formation and maintenance and the ways that society impacts the body, which is an idea not often considered in medical studies. Such influences are not trivial and are likely to become exposed as a transplanted head interacts with its new body. Society and social circumstances, usually expressed in the form of expectations or reactions, train the body and the mind to manage those social "things" to survive, cope, or prosper. It is my intention here to explore these dynamics in context of the neurological and psychological processes involved in responding to the social environment, all of which will be disrupted by a head transplant.

Acknowledgments

Special thanks go to John Barker who, unlike me, is taking retirement seriously. I appreciate your readings of some of the "not boring" medical sections and avoiding the "boring" sociological and bioethical parts. Fortunately for us over the years and me in this moment, our experiences of what is boring are complementary.

We did not cover spinal cord anatomy in sociology school, or perhaps I was absent that day, hence my need for tutoring all things neurological. For that I turned to my friend Amanda Nix for guidance and patience as I tried to learn the basics of how the spinal cord works. Thanks, Amanda. Also, my appreciation goes to two consulting neurologists for their opinions on head transplantation. One of these physicians I have not met but graciously gave me time via email to answer a few questions. The other is an old friend from high school, who needless to say, did better in the high school biology class we took together. Since I forgot to ask if I could mention their names here, we will just refer to them as neurologists of high esteem who helped me out. Thanks to both of you.

As always, all hail the great and powerful Abby Shapiro, my wife, best friend, and keeper of all things grammatical. Just as I was absent on neurology day in graduate school, so must I have missed the class on maintaining verb tense continuity in manuscripts. She, however, did attend that lecture, developing a keen eye for catching my seemingly pathological tendency for mixing present and past tenses over the course of several paragraphs. Thanks, Abby, yet again.

Contents

1

An Introduction to Transplanting Human Heads

Chapter Summary

Despite media coverage of one surgeon's plan to conduct the first human head transplant, also known as body-head transplantation or BHT, the idea of placing the head of a dying person onto the body of brain-dead but otherwise healthy donor is not well recognized and has been viewed as pseudo-science. This chapter introduces the concept of body-head transplantation and its intended purpose and strategy. While many, though not all, of the surgical techniques needed to conduct this experiment are in place, critics from several disciplines have expressed grave apprehensions about the ethical merits of BHT, criticisms that have dogged the proposal and denounced its legitimacy since its inception. After outlining the major concerns that critics have raised, the chapter provides a pragmatic conceptual framework for evaluating these ethical assessments.

*   *   *

With each passing year the pace at which the human imagination evolves from fantasy to practicality grows faster. While much remains unknown, the mysteries of life are dwindling. In the 21st Century, we are of the mind that there are no limits to our creativity. We can make or do almost anything, and when confronted with hurdles, we strive, and often succeed, to find a way to overcome them. That which we can dream, we can do.

Indeed, what was unimaginable or impossible a mere two centuries ago are unexceptional today. Not that long ago rapid transit, space flight, instant global communication, and holding access to the knowledge of the world in a small handheld device were unthinkable. Our ancestors could not imagine the post-modern relative relaxation of social perceptions of race, ethnicity, and gender and the ascension of individualism that has redefined marriage and family life. Nor could they foresee the epidemic of existential crises that seems to penetrate every aspect of daily living and begats a question that then was inconceivable: "Who am I?" The most farsighted of our forebears would likely perceive much of today's technology as frightening or alien to the human life they experienced and understood.

To that point, compared to the turn of the 20th Century, today's health care practices would be considered miraculous. In 1900, life expectancy in the United States was under 50 years. Now people in this country can expect to live about half again longer, approaching 80 for women and 74 for men. In other countries, gains have been more dramatic. People born in Japan and Switzerland, for example, can expect to reach their 85th birthday. Because of a profusion of civic improvements in public health and advances in the control of infectious diseases that spurred longer lives in the early 1900s, the social admiration of western medicine, especially compared to that of the 19th Century, has soared. As the social group intervening between the populace and the medical institution, physicians are now bestowed with great authority and near reverence, and they are expected to cure or treat anything and everything. While humans have always understood their impending mortality, we now have the means, and accompanying expectations, to delay our eventual ends longer than any time in the past. To do that, we entrust doctors and their array of machines, chemicals, and ingenuity with the task of keeping us alive.

Of all the recent medical innovations, one that would surely amaze the scientists and physicians in "olden times" would have been the ability to replace damaged or diseased organs and tissues with healthy ones taken from other people. History tells us that western scientists investigated transplantation as far back as the 17th Century, an exploration that carried into the 1800s. Although these early transplants did not accomplish much in terms of organ functionality or extending life, the advances in surgical techniques gained through the failures of these pioneer surgeons provided the foundation upon which the techniques required to remove and reconnect various body parts were set; if nothing else, these seemingly minor surgical achievements foreshadowed the accomplishments that laid just beyond the horizon.

Not long ago, we asked, "What parts of the body can we transplant?" Now we are asking, "What cannot be transplanted?" Solid organ transplants have become commonplace. Transplanting kidneys, hearts, lungs, and corneas, among other organs and tissues, is everyday practice. Experiments transplanting legs and arms are taking place in various hospitals as are larynx, pancreatic islets, and thymus glands. Other more "exotic" body parts have also been transplanted as well. Penises and scrotums, uteri, and ovaries have been transplanted at various medical centers throughout the world, as have hands and faces. What is left to transplant? A casual glance at the human body reveals the answer: a human head.

Transplanting a Human Head

Not long after the conclusion of World War II, a revolution in the history of medicine was about to be born, birthed off the trials and disappointments of previous generations of physicians who hoped to save lives by transplanting organs. This revolution began a movement in medicine perhaps rivaled only in scope by the discovery of "germ theory" that preceded it and genetics medicine that would follow. Rather than one huge explosive event, the transplant medicine revolution occurred in three distinct phases or waves, each boundaried by major discoveries that transformed the nature of transplantation and expanded its accessibility and the ability to save or genuinely improve people's lives.

There are two primary hurdles that must be overcome to perform successful transplants. First, surgeons must master techniques that allow them to remove and reattach the organ or tissue so that it will continue to function. There are many such procedures. Doctors must learn the proper means for cutting, suturing, preserving the severed tissues, and matching the donor's blood type and other characteristics to that of the recipient.

The mechanics of transplantation were largely worked out by the 1950s, but the second impediment to transplant success seemed insurmountable. The problem was that the body recognizes donated tissues as a foreign invader that must be resisted. The body attacks it, as it would a virus, in a process known a tissue rejection. Although today tissue rejection is common knowledge, prior to the 1950s, the details of the immune system were still largely undiscovered. More was learned and knowledge became increasingly sophisticated so that by the early 1950s, research physicians had a basic idea for ways to lower the chances of tissue rejection. On that idea, the onset of modern transplant medicine began.

The First Wave of modern transplantation ushered in the era of solid organ transplantation and began with Joseph Murray's renal transplant between identical twin brothers in 1954. The recipient lived eight years after his surgery, a feat considered astonishing at the time and earned Murray the Nobel Prize in 1990. Five years later in New Orleans, doctors transplanted a kidney donated by a man to his fraternal twin brother who lived 25 years after the surgery. These two marvels confirmed the long anticipated feasibility of transplanting solid organs; however, these and other transplants among siblings in France and elsewhere did not completely solve the problem of a recipient's body rejecting a donated organ. The First Wave surgeons proved that genetically related donors and recipients could share organs with a high probability of a positive outcome, but limiting donation to close family members did not permit surgeons to fill the ever-increasing demand for life-saving transplants. Tissue rejection remained the central barrier. How could the body's immune system be constrained so that foreign tissues would be accepted?

The second revolution in transplant medicine began with the discovery of pharmaceutical means to minimize rejection. The Second Wave followed dramatic improvements in immunosuppression that began to erode the limitations of restricting donations to close family members. The revolutionary discovery in the early 1960s that azathioprine, actinomycin C, and prednisone convincingly and relatively safely suppressed the immune system so that non-related individuals could serve as organ donors. With these drugs, the risk of rejecting foreign tissues was significantly reduced to acceptable levels. The impact of this new immunosuppressing "cocktail" was immediate and profound, and the number of transplants and the rate of their success increased sharply. It also amplified the pool of potential donors, making possible life-saving treatments and allowing for the transplantation of other and more immediately life-saving organs such as the liver, heart, and lungs, all of which are now everyday occurrences throughout the world.

The impact on human life of the Second Wave is undeniable. Organ transplants now occur on all continents, except Antarctica, and in 2020 saved just under 130,000 lives (WHO 2020). In the United States, a record 42,887 life-saving transplants were performed in 2021 (U.S. Department of Health and Human Services 2023). To date, the number of people whose lives have been saved or made easier by Wave Two inventions is well into the millions.

Despite the success of these early medications in suppressing the immune system and preventing organ rejection, the same dosages used to successfully prevent organ rejection were not effective in preventing rejection of other tissues, especially skin. The drugs that minimized rejecting a solid organ proved insufficient for composite tissues, as proven by an Ecuadorian surgical team in 1964 that conducted the first known hand transplantation. While the surgery resulted in mechanical functionality, the post-operative immunotherapy relied on drugs used to prevent rejection in kidney transplants. The hand was rejected within two weeks and had to be removed.

Each tissue in our body (such as kidney, liver, heart, muscle, bone, tendon, and skin) possesses a different degree of "rejection-ability" or immunogenicity. Skin's high immunogenicity is due to one of its primary functions, which is to serve as a barrier that prevents foreign materials from the surrounding environment from entering the body. For this reason, organ transplants had been performed successfully for many years, and yet transplantation of tissues containing skin (hand, foot, and face, for instance) was not possible until relatively recently.

From the late 1970s through the mid-1990s, several research teams continued to test the effectiveness of different combinations of immunosuppression medications being developed for use in transplantation in a variety of different tissues, including skin, especially the hand. In clinical trials that used dissimilar types of tissues including nerves, tendons, muscles, bones and joints, larynx, abdominal walls, tongues, and uteri,[1] outcomes were generally positive; however, in separate animal experiments, these new medications were still not able to prevent rejection.

In the late 1990s, a team of researchers at the University of Louisville stumbled upon a combination of immunosuppression medications that effectively prevented limb rejection in a pig limb model. The researchers were comparing a group of animals treated with medications delivered locally via implanted pumps, with a control group that received a 3-drug combination (tacrolimus/mycophenolate mofetil/corticosteroid), which at the time was considered the gold standard in clinical kidney transplantation (Ren et al. 2000; Ustuner et al. 2000). In the study, the pumps, which were administering a single dose of cyclosporine to the experimental group, malfunctioned; however, in the control group receiving the 3-drug combination, the limbs experienced little or no skin rejection.

Based on these findings, the Louisville team applied for and was granted permission from the hospital's ethics review board to conduct 10 hand transplants in clinical trials. The Louisville team's data were presented and published (Shirbacheh et al. 1998; Jones et al. 1999) and in 1998 and 1999 teams in Lyon, France (Dubernard et al. 1999), Louisville, Kentucky (Jones et al. 2000), and Guangzhou, China performed the first successful human hand transplants (Francois et al. 2000) using this same three drug combination (tacrolimus/mycophenolate mofetil/corticosteroid).

Hitting on the right dosage of tacrolimus/mycophenolate mofetil/ corticosteroids solved the problem of transplanting skin and had an immediate impact on transplantation medicine. This discovery opened the door for the first successful hand transplants and the many that followed. As of 2020, about 150 hands have been transplanted on 100 recipients worldwide (Alberti and Hoyle 2021), and while acute rejections have occurred in 85 percent of those cases within the first year, swift dosage

---

[1] To learn more about the evolution of composite tissue allotransplantation, please see Bain (2000) about peripheral nerve transplantation, Mackinnon et al. (2001) and Mackinnon (1996) on transplanting neuromuscular and nerve tissues, Guimberteau et al. (1992) regarding tendons, Jones et al. (1998) for muscles, Hofmann and Kirschner (2000) for bone and joint transplantation, Strome et al. (2001) regarding laryngeal transplants, Levi et al. (2003) for abdominal walls, Bhattacharya (2003) about the first tongue transplantation, Fageeh et al. (2002) uterus, and for hand transplantation, Daniel et al. (1986), Stark et al. (1987) and Hovius et al. (1992).

adjustments were able to reverse them. Furthermore, a review study in 2019 found no reports of graft-versus host disease or chronic rejection among recipients of new hands (Reece and Ackah 2019).

The Significance of Third Wave Transplantation

The loss of a hand is catastrophic, leaving a patient with an impairment that can have devastating effects on the person's ability to complete everyday tasks commonly taken for granted. Those who have lost one or both hands often experience economic dependence, difficulties with self-care, and overwhelming emotional injuries to their self-image and self-esteem (Errico et al. 2012). Amputees often say that they cope with an often unfriendly social environment by withdrawing, feeling embarrassed at their appearance and perceived shortcomings in comparison with social norms of ability and attractiveness. Depression is a common conclusion to the mismatch between their physical bodies and social expectations of what a body should look like and do.

Improved psychological well-being, therefore, should follow a hand transplant, and that is usually the case. Studies of patients who received transplanted hands at the various transplant centers around the world show that quality of life does indeed get better for a large majority. Self-esteem improves, and most patients incorporate their new appendage(s) into their self-image (Schuind et al. 2006). There have been exceptions, however. The first recipient in the Third Wave of transplantation history was unable to integrate his new hand into his corporeal self and reportedly tried to remove it himself before reaching doctors who surgically removed the hand. In a sense, it is fortunate that this incident happened early in modern hand transplantation history because it called attention to the psychological wherewithal of potential transplantees and showed that a pre-operative psychological evaluation was as important to surgical success as the physical examination.

Not long after skin rejection was overcome and surgeons demonstrated that hands could be transplanted, the human face became the next surgical target. The face posed more intricate clinical and ethical barriers than hands, however. Since face transplantation, as well as that of hands, is a life-enhancing procedure, as opposed to lifesaving, the strongest arguments came from those who believed that the toxicity of life-long immunosuppression was too risky for people whose only medical problem was severe disfigurement. On the surface, that seems like a reasonable position; however, the impact of severe facial disfigurement runs deeper than mere appearance.

The social meanings of the face, once described as "the epicenter of social identity" (Furr 2014), make it perhaps the most socially important part of the body. It bears clues not only to our physical identity and attractiveness, as perceived by cultural standards, but also to our social class, age, ethnicity, and gender. People whose facial features have been burned off or grossly deformed through trauma or surgery for cancer usually confront social and psychological complications incomprehensible to people who are not disfigured. Typically, people with severe facial deformities become "shut-ins" whose life spaces have shrunk because of the way they are treated in public settings. Strangers often respond to them with startles, invasive questions, or insulting and demeaning comments about their appearance. Such encounters have conditioned them to avoid contact with other people. Consequently, disfigured people are at risk for depression, anxiety, and suicide.

The loss of physical identity and body conformity is only a part of the misfortunes severely disfigured individuals must tackle. Functional limitations also contribute to the hardships they endure. If one's eyelids are severely burned and scarred, blinking is impossible, and eyes must be regularly cleansed with saline drops to remove particles and keep the corneas moist. No lips or having only a partial jaw impair talking and masticating, and the loss of a nose makes breathing difficult. Many disfigured persons have chronic facial pain; some may be blind. While true that face transplantation is not essential to save lives, the quality of life improvements offered by a new face are truly extraordinary. Facial transplantation is not about making a person attractive; it restores a person's ability to live more comfortably and eases the emotional pain of having veered so far off the norm of expected appearance that other people become afraid at their sight and dehumanize them to the point of believing they are demonic beings (Furr 2014).

Face transplant patients have indeed found relief of these physical and social injuries. After their surgeries, patients generally report improved social relationships and quality of life and feel that their new appearance is acceptable not only to themselves but to others. Anxiety about their appearance declined significantly after their surgeries, and many feel more confident in public settings and are less afraid that others will respond to them with hostility, rudeness, or inappropriateness (Nizzi et al. 2017). Quality of life, judged by improved facial functioning, also contributes to improvements in the overall mental health of face recipients (Pomahac et al. 2014).

More complex than hand transplantation, evidence suggests that face transplants carry more risk. Of the 48 recipients of new facial tissue to date, we have outcome data on 39, and of these cases 90 per cent have experienced acute tissue rejection, a slightly higher risk than among hand transplantees. As of 2021, five individuals dealt with chronic rejection, compared to no hand transplantees. More tragically, eight patients have died. Their causes of death have been reported as malignancy (in two instances), noncompliance, infections, sepsis, suicide, respiratory failure, and hepatocellular carcinoma, a common form of liver cancer (Diep et al. 2021). It is unclear to what extent these deaths were connected to the surgery itself, the ensuing immunosuppression therapy, or the psychological aftermath, but the first recipient, whose death from a malignancy about 11 years after surgery, is known to have been a consequence of complications from the immunosuppression medications.

As a sidenote, despite its potential severity, rejection of skin in a CTA differs from rejection of solid organs. Most importantly, skin rejection, expressed as a rash, can be detected early, making it possible to immediately adjust the immunosuppression medications and eliminate the rejection episode. Whereas in organ transplants, the time to detect rejection is much longer and can potentially have a more injurious aftermath. A second immunological advantage to CTA is the people who receive an organ are ill with the disease that caused them to need the organ. If they have not fully recovered from that illness, the rejection can compound it, making them more ill. People who receive hands or a face are not necessarily sick, and the rejection in most cases is more of a solitary and containable incident than that which occurs with solid organ transplantees.

Every year yields improvements to both hand and face transplant processes. We know more about immunosuppression, and research continues at full speed to identify other means of tolerance as alternative to the drugs that must be taken for the rest of the recipient's life and that carry risks due to their own toxicity. Lifelong immunosuppression exposes patients to various illnesses because, to state the obvious, their immune system is compromised. The heightened risk of transplant patients acquiring COVID-19 serves as a stark reminder that we have far to go in making transplantation of any type, solid organ or CTA, a totally safe treatment option. These risks of lifelong immunosuppression continue to be a main ethical argument against hand and face transplant since these are life-enhancing procedures and not typically life-saving.

Prior to the first face transplant, the surgical research needed to ensure its success had been completed to the point that physicians were highly confident that the surgery would work. Surgically removing the donor's healthy face and the recipient's damaged facial tissues had been fully rehearsed and tested, and the immunosuppression protocols were likewise established and approved. In fact, addressing the ethics of giving someone a new face proved to be the last barrier to the first attempt.

As ought to happen when a revolutionary new medical procedure is conceived, a rousing debate on the "should" question (as in should the surgery be conducted) was carried out and presented in scholarly journals. Articles both for and against face transplantation made strong appeals for their case. Since surgical techniques were essentially off the table because they were grounded in sound evidence and established practice, the main themes against transplanting a face centered around three points. First, like the critique of hand transplants, exposing patients to the risks incumbent in life-long immunosuppression did not warrant the benefits of a life-enhancing operation. These risks, including cancer and opportunistic infections, did not balance the rewards that a new face offered. Second, questions were raised about identity formation and psychological coping as patients adjusted to seeing a different person in the mirror after the transplant. Hand transplant patients also needed to adjust to receiving such a personal body part, but studies suggested that most recipients accepted their new hands without psychological distress and integrated their new appendage into the physical self-image and identity.

And third, strong positions were held, especially in Britain, that disfigured people did not have a problem — the problem was society and its bigotry and stereotyping tendencies towards people with facial deformities (Hughes 1998; 2019; Furr et al. 2017). The resolution of the problem of how people in general related to those with disfigurement was not medical but in a change in public attitudes that promoted greater tolerance of physical differences. While the latter point is indeed correct, it is also an ideal. Social change of this type evolves rather slowly and is not a simple task. As such, reversing eons-old biases against facial disfigurement was not an immediate solution to the prejudice and discrimination disfigured people endure. Plus, this argument does not account for functional limitations that often accompany severe disfigurement. Not all problems facing disfigured people are aesthetic.

These were compelling arguments against transplanting a human face. What is of interest here is that all these opinions could be challenged as patriarchal in form; that is, medical authorities and bioethicists were telling patients what was good for them without dialog between them and the people who actually had the experience of being disfigured. Two studies (Barker et al. 2006; Brouha et al. 2006) sought to remedy that deficit in the debate by asking two groups of people, one with severe facial disfigurement and the other solid organ transplantees already on immunosuppression, to reflect on their experiences and what they believed was in their best interests in terms of immunosuppression and quality of life.

The studies sought to determine how much life the people in these two groups would be willing to lose to get a new face or to continue to receive immunosuppression. The results revealed that disfigured people would indeed trade-off a few years of expected life with life-long immunosuppression in exchange for the benefits of a new face. Similarly, kidney transplant patients held few regrets in taking the immunosuppression medications that could potentially shorten their lives and willingly made that trade for the benefits of their new organ. The consequences of disfigurement were of such magnitude that the risks of immunosuppression were acceptable when offered the benefits of restoring facial appearance and function (Barker et al. 2006). By satisfying some of the ethical discontent about face transplants, these studies helped set the stage for the promotion of face transplantation and its proliferation.

The pre-surgery ethical research on face transplantation was large, disciplined, and eventually evidence based. While not all the moral answers could be derived empirically, researchers provided sufficient ethical grounds on which to proceed. The ethical "pros" appeared to outweigh the "cons", and IRB proposals were developed and approved, and face transplants became a reality and while not routine, they are no longer newsworthy events.

We have now come to a turning point in the life-saving versus life-enhancing debate about CTAs. We can no longer say that composite tissue allotransplantations are solely life-enhancing because a novel CTA concept waits in the wings and bides its time before its fantasy evolves into reality. This one could be the most complex surgery ever attempted and its attending ethical baggage makes the discussions on hands and face transplants appear as bioethical exhibitions leading to the main event. The time has come to talk about transplanting a human head.

Are Head Transplants the Next CTA?

Given all the parts of the body that have been transplanted thus far, placing a person's head and brain onto another's body is perhaps the final frontier of transplant medicine. Head transplantation, however, is no ordinary surgery. To transfer a head to another body involves multifaceted medical challenges as well as complicated ethical and existential dilemmas that have only recently begun to be debated. Because of the surgical and moral intricacies, many physicians and ethicists do not agree that transplanting a head is the logical continuation, and perhaps conclusion, of Wave Three transplantation. Could it be the beginning of a Fourth Wave, or is it so dangerous and oppositional to established social norms that it should not be attempted at all?

If head transplantation is the next milestone, it should follow the same developmental research process as hands and face, both in terms of surgical development and bioethics sophistication and appraisal. Many have stated that this has not been the case, contending that proponents of transplanting heads are prematurely promoting the procedure and trying to establish its legitimacy without following medical and ethical due process of research or addressing procedural and bioethical criticism. Herein lies the divide. If we consider the surgical and ethical explorations prior to the first hand and face transplants as the template for technological innovation and adoption in medicine, we must ask if head transplantation has satisfied that model.

Is "Head Transplantation" the Right Name?

First, we must clarify an appropriate name for this procedure. While "head transplantation" is the usual term given to transferring a healthy head from a dying body to a brain-dead but otherwise well body, it may not be correct because it is not clear what part of the body is being transplanted.

Some contend that the body is the recipient and is receiving a transplanted head, thus the term "head transplant" is indeed correct. Claims that the body is the host are based on the location of the immune system, which primarily resides outside the head. The idea is that the body would reject the head. In this framework and using neurological and immunological conceptions of the notion of selfhood, the body is the self, and the head is the non-self, making this surgery truly a head transplant.

Others, however, argue that the head is the recipient, and the body is the donor. In this model, "body transplant" is the appropriate label. Those who maintain this side of the debate use several arguments to make their points. First, they posit that the term "head transplant" is confusing because it is inconsistent with the traditional descriptions used in transplantation. Typically, the unhealthy organ that is being replaced by a donated healthy one names the procedure. For example, a person who undergoes a liver transplant has a damaged or diseased liver that must be removed. The head, however, is not the part of the body that is diseased or injured and in need of replacement; it is an impaired body that is being exchanged for one that is healthy. Second, this position acknowledges that immune functions occur in both the body and head. The face could reject the body just as easily as the body might reject the face and head. Lastly, the self, as defined in Western psychology rather than in mmunology, resides in the head in the form of memories, personality, identity formation, and cognitions. Therefore, from this perspective, the donated body is being transplanted, and the head is the recipient, and the procedure should be known as a body transplant.

Although the points of contention made by both sides of this question are important and significant in the grand scheme of understanding the complexities of the experiment, they sit on the sidelines of the aims of the present analysis and will, essentially, be left unattended. Nevertheless, both positions raise important questions for how the experiment would (or possibly will) play out and influence the ethical discussion of connecting a head from one body to that of another. Still, neither term is convincingly accurate on its own. Because we must call the surgery something, and its occasionally used semi-official label of cephalosomatic anastomosis is cumbersome and also not completely correct,[2] we will refer to it as body-head transplantation or BHT. This designation has been used in the scholarly literature and satisfactorily placates both sides of the argument by including the key points of both in determining the experiment's likelihood of success and constructing a paradigm for assessing its ethical merits.

Why Body-Head Transplantation?

BHT is a proposed surgery to save the life of a dying person by separating the head from an incurably diseased body and then reattaching it to a brain-dead, but healthy body. The goal of a BHT, unlike face and hands transplants, is to sustain the life of people who suffer from a terminal disease but whose head and brain remain healthy. Ideally, this surgery would provide a lifesaving treatment for conditions such as amyotrophic lateral sclerosis (ALS or Lou Gehrig's disease) or end-stage cancer, among many other life-threatening conditions. BHT would be a last-resort medical option employed after all other known treatments have been exhausted or a cure is unknown.

Despite its potential to save lives, the possibility of replacing an incurably ill body with a healthy one tests not only our surgical limits but calls into question the social and psychological boundaries of physical life and alters what we recognize life to be. Neither hand nor face transplantation raised the depth of medical and ethical inquiry that has BHT, and the discourse only promises to intensify as the call to conduct the surgery persists.

Before moving forward to the ethical matters, however, we must understand what BHT entails. Knowledge of the basic steps of BHT are accessible to those without medical training and will be reviewed here. The details, however, are another matter. BHT will involve not only the highest level of surgical expertise, but also the highest degree of skill in biochemistry, neurology, immunology, orthopedics, cardiology, anesthesiology, and nursing. The medical specifications of BHT are not necessary for our story at this point, and a general outline of the procedure suffices to open the discussion on the ethical merits of the surgery. Medical details, when they are needed to sort out ethical questions, will be introduced in later chapters.

---

[2] Anastomosis is defined as a surgical connection between two structures. The term, however, usually refers to connecting tubular or body channels. According to the Cleveland Clinic's definition, surgeons create an anastomosis after removing or bypassing blood vessels or parts of the intestines or following the removal or transplantation of an organ that was joined or linked to a channel. It is not a term that typically represents a transplant procedure.

BHT as an Old Idea Reinvented

Nothing exists in a social or cultural vacuum, and everything has a history. The present unfolds from context: new ideas are born from pre-existing ones. The current interest in BHT is no exception. The idea of body and head transplantation has a long but rather obscure past. Though the medical history of CTA and BHT dates back just over 100 years, the notion has existed for over two millennia. The idea of composite tissue allotransplantation, though not known as such at the time, was perhaps first recorded over 2,000 years ago, not in the material actuality of medicine, but in the immaterial reality of religious mythology. We cannot know for sure if they invented the concept, but the authors of sacred Hindu texts may have provided the first written accounts of moving heads from one body to another.

The ancient Vedic texts, the Upanishads, tell us of the symbolic and medicinal importance attached to the human head. In these scriptures we learn that deities and happiness of the soul dwell within the head: one honors the gods with mental activities (Brown 1921). Perhaps these symbolic interpretations explain why when portrayed in human form Brahman, the Hindu god of Creation, has four heads. To Hindus, Brahman represents the highest of universal knowledge and the infinity of existence. Brahman is the soul, power, and wisdom of all that endures, the ultimate truth, and the supreme ruler. Multiple heads can symbolize many things: the five elements (earth, wind, fire, water, and ether) or the different elements of life such as creation, preservation, and the next cycle of creation.

Brahman is not the only Hindu deity with multiple or transplanted heads (Harikrishnan 2019). Lord Ganesha, who is among the most well-known and beloved gods within the Hindu pantheon, has a human body and an elephant head, a conjoining of significant moral import. In a misguided fight, Ganesha's father, Shiva, killed his son by severing his head. Feeling remorseful, Shiva tells Ganesha's mother that even he is unable to restore his son's head to its body but instructs his fellows to go into the forest and bring back the head of the first creature they encounter that is separated from its mother. The first they came across is an elephant calf that had become lost from its parent and wondered alone in the world. Shiva attaches the elephant's head to Ganesha's body, restoring his life and endowing him with important duties and honors. The story of Ganesha teaches us to control egoism and align ourselves with the universal truths of Creation. Believers pray to Ganesha to remove the barriers that stand in the way of their fulfilment and to grant success and prosperity.

Other Hindu deities exist as xenotransplantations (transplanting or grafting tissues from a nonhuman animal source to a human recipient).

One of the Creators, Daksha, has a human body and the head of a goat. Tumburu, the great musician, has the head of a horse, and Kamadhenu, the sacred mother of all cows, has a human female head and a bovine body. The lives of these gods are complex, and their animal heads or bodies give them unique powers of wisdom, providing models and guidance to help mortals manage their daily affairs and make sense of themselves and the world. In religious literature, transplantations perhaps are metaphors for righteousness, the divinity of existence, and the restoration of life.

Religion was not the only cultural space in which proto-CTAs occurred in ancient India. Sushruta, who lived about 2,000 years ago most likely in Varanasi, developed advanced facial reconstruction techniques, especially for rhinoplasty. Among other medical achievements, he was known for treating men whose noses had been severed as punishment for a criminal offense by using skin grafts from their foreheads to form what is often called the "Indian flap", a method that differs little from today's practices. Though not formally a transplant, Sushruta was among the first to document the possibility of taking human tissue from one site and placing it elsewhere. In terms of intellectual evolution, perhaps Sushruta's was a revolutionary thought that must occur before thinking of moving a body part from one organism to another, an example of a medical walk before the run.

Multiple tissue transplants were not unknown in Christian antiquities either. Syrian Christian twin brothers Cosmas and Damian were wandering third century physicians with a reputation for relieving mundane ailments as well as affecting miraculous cures of the most serious illnesses. Their fame exploded when it became known that they plied their medical talents without charging a fee, and a cult of worship was created by devotees. By the mid-5th Century, their followers were widespread and founded churches and holy sites in the brothers' honor throughout the eastern Christian world. One of their wonders became immortalized a thousand years later when it was recorded in a 13th Century collection of hagiographies and later commemorated in a famous painting by Matteo di Pacino (ca. 1370). The painting provides a visual image of the brothers' most famous medical marvel: the Miracle of the Black Leg. The "miracle" occurred when they transplanted the leg of an Ethiopian or Moor (there are several versions of the story, but the donor is always dark skinned) onto the body of a White church verger who was dying from "cancer", which was probably gangrene or perhaps a carcinoma (Duran 2018; Jović and Theologou 2015). The verger survived, and Cosmas and Damian, who were eventually martyred during Roman Emperor Diocletian's persecution of Christians, were later venerated as the patron saints of surgeons and pharmacists. The racial connotations of the myth and the painting are a separate story (see Goodrich 2020), but the idea that body parts can be harvested and used to save the dying may have begun, in the western world at least, with this legend.

By the late 18th Century and throughout the 19th, western physicians were experimenting with composite tissue transplantations of various types, some of which, in modern thinking, were peculiar first steps, though in their own socio-historical context they probably made scientific sense. Scottish surgeon John Hunter, for example, transplanted a human tooth onto a rooster's crest, among other similarly odd experiments. Even the famed early endocrinologist and spinal cord researcher Charles-Édourd Brown-Séquard transplanted tails of rats and cats onto rooster crests (Duran 2018). Physicians of the Industrial Age were conducting critically significant transplants as well. Despite reports that showed they did not work, attempts to transplant whole ears and noses persisted in the 1800s. By the 1870s, physicians realized that autografts (transplanting tissue from one place to another on the same person's body) were superior to homografts (transplanting tissue from one person to another) (Goldman 1987), but meaningful success proved elusive, except that techniques for tissue "harvesting" and replanting improved, albeit quite slowly.

Despite the persistent failures, experiments to improve the mechanics and the implements necessary to transplant organs and composite tissues continued through the 19th Century into the 20th. Although a complete understanding of the immune system did not exist at the turn of the last century, techniques became increasingly better with each passing year. Effectively stitching together blood vessels and maintaining blood supplies, for example, were unresolved problems in 1900, and it was to that end that the first recorded removal and reattachment of an animal head was conducted in 1908.

In collaboration with French surgeon Alexis Carrel, American physiologist Charles Guthrie was earnestly trying to advance revascularization. The two tested different methods of closing wounds to blood vessels and found that tiny, near invisible sutures sewn into the ends of vessels worked best if those ends were turned back like a cuff and stitched together. The work had dramatic implications for injury repair and general surgery, but also for the hopes of transplanting organs. Carrel was known to have transplanted organs in dogs, proving that the procedural mechanics needed to perform such surgeries existed. But Guthrie took experimenting with the new suturing methods a step further: he removed a dog's head and surgically connected it to the body of a recipient canine (Guthrie 1912). The dog only lived a few hours but progressed Carrel and Guthrie's understanding of revascularization. For his work in suturing blood vessels and advancing organ transplantation, Carrel, but not Guthrie, received the 1912 Nobel Prize for medicine.

That award was not only significant in terms of recognizing Carrel's advances in revascularization, but it also marked the first ethical challenge to body and head transplantation. The 1912 award would become the subject of controversy when some contended that the credit for the research should have gone to Guthrie. It is believed by some that Guthrie's head transplant experiments, unlike Carrel's organ transplants which were championed, likely influenced the Nobel committee's decision to exclude the American from the award because the committee members were disgusted by those experiments (Stephenson et al. 2001). Apparently, Guthrie had crossed a moral line.

The center of early BHT would then shift eastward to Russia and the Soviet Union. The decades following Carrel's Nobel were not kind to that country. The Tsar's tyranny, World War I, revolution, civil war, economic privations, famine, and the German invasion of 1941 created traumatic hardships for people throughout the Soviet Union, and the nation's medical infrastructure was unable to relieve the suffering. Amid those horrible conditions, Russian scientist and organ transplant pioneer Vladimir Demikhov hit upon the notion of creating a comprehensive tissue bank to store organs for transplantation. He envisioned that transplants could save countless lives from the pain and distress in which his people had been emersed. An expert in heart disease, he developed several surgical techniques for transplanting hearts and lungs and was beginning to receive worldwide attention in the two decades following World War II. While Demikhov's innovations focused primarily on developing surgical procedures to transplant hearts, the Soviet scientist, like Guthrie did earlier, grafted a canine's head onto another dog's body to test techniques and approaches to organ transplantation. As with Guthrie's, Demikhov's dogs did not live beyond a few hours or days, but while the American's experiment was not widely noticed outside his own medical and scientific network and the Nobel review committee, Demikhov's two-headed dog was documented on film, which was surprisingly released to the world. In the footage, which lasted only a few minutes, viewers can see both heads lapping up milk and panting. It must have been a macabre sight, but in addition to emotional reactions most people would have at seeing a freakish beast and the product of a real-life "mad scientist", many American scientists and politicos also interpreted Demikhov's achievement with Cold War dread. Here was a case of Soviet science working beyond the ability of the western democracies and capitalism. If the Soviets were putting their two-headed dog on film, what must they be hiding (Schillace 2022)? Demikhov's experiment caused a political stir in the United States about Soviet scientific progress during the time of Sputnik, nuclear proliferation, and international competition for cultural, political, and military hegemony. Experimental medicine now had nationalistic and even imperialistic implications, foreshadowing an element of BHT that would occur 50 years later and will be a part of our story[3].

A patriotic reaction would manifest not long after the film recording was released and reported by the press. Within a day of reading of Demikhov's two-headed dog in the local (Cape Town, South Africa) newspaper, Christiaan Barnard, the renowned heart surgeon, conducted the same operation (Hoffenberg 2001). A colleague who was present when Barnard learned of the Russian's feat stated that Barnard became incensed and stormed out of the room saying, "anything those Russians can do, we can do, too" (Konstantinov 2009:456). By the end of the day, Barnard created his own two-headed dog, which survived a few days. As Hoffenberg (2002:1478) would later assert, "there was no clear purpose to this [Barnard's head graft] other than to show his technical virtuosity." The experiment furthered Barnard's local celebrity, and students at his medical college made a papier mâché dog with two heads and paraded it around campus in recognition of his achievement.

The center of BHT experimentation turned from the Soviet Union to the United States in the late 1960s and 1970s. Neurosurgeon Robert White led an American team in Cleveland that performed head transplants on monkeys (White et al. 1971). Benefitting from newly developed immunosuppression medications used in solid organ transplants and working with rhesus monkeys, White surgically removed the head of one monkey and connected it to the body of a medically decapitated other. He claimed success and reported postoperative restoration of smell, taste, hearing, and motor function of the face in the transplanted heads. The high dose of immunosuppression required to prevent the head from rejecting, however, caused the monkeys to die within nine days after the transplant (McCrone 2003).

Despite this remarkable neurosurgical achievement, White's head transplant research received extensive criticism, particularly from animal rights activists who called him barbaric, torturous, and cruel (Schillace 2022). The monkeys remained paralyzed from the neck down and experienced great pain, according to Jerry Silver who was present during White's 1970 experiment. Silver later told CBS News, "I remember that the head would wake up, the facial expressions looked like terrible pain and confusion and anxiety in the animal. It was just awful. I don't think

---

[3] Again in 1968, Demikhov transplanted another puppy's head onto the neck of a more mature canine. The animal died after 38 days. Twenty years later, the dog's taxidermied remains were donated to the Museum of History of Medicine in Riga, Latvia, then part of the Soviet Union. From 2011–2013, the stuffed two-headed dog was displayed at various exhibits in Germany before returning to Latvia as rejoin the museum's permanent collection (O'Carroll 2013).

it should ever be done again" (Elliot 2013). White himself was branded "Dr. Butcher" (Bennet 1995), and that label would remain attached to his reputation throughout his career. The effect this moral reproach had on White was captured in his own words, "it is now possible to consider adapting the head-transplant technique to humans. Whether such dramatic procedures will ever be justified in the human area must wait not only upon the continued advance of medical science but more appropriately the moral and social justification of such undertakings ..." (White et al. 1971:604).

Since White's experiments, few animal studies have been conducted, and generally these were not intended to test the feasibility of head transplants. For example, Hirabayashi et al. (1988) and Sugawara and colleagues (1999) transplanted infant rat heads to study the regulation of craniofacial bone growth, and Atsushi Niu's team (2002) transplanted rat heads to study brain function following long periods of ischemia. In fact, none of the previous BHTs or head grafts were conducted solely to test the feasibility of transplanting a head. Even White's monkey transplants had a hidden agenda. White, who was a devout Catholic, wanted to transplant a human head to identify the location of the soul (Schillace 2022). He believed it was situated in the brain but felt that the only way to prove it was to place a brain-containing head onto another body to see if the person inside the head, that is, the same soul, woke up afterwards. In fairness, White was also convinced that BHT would save lives, so his work shared religious and lifesaving motivations.

It was not until several decades following White's experiments that a research team experimented with BHT for the sole purpose of transplanting a head. Xiao-Ping Ren's group in China, working in conjunction with Italian neurologist Sergio Canavero, has transplanted hundreds of mouse heads with the intended purpose of verifying Canavero's theoretical model for transplanting a human head. While Ren's mice have not lived long (measured in hours and days), he reports achieving some level of success but falling short of full recovery in terms of complete spinal cord connection, immunological suppression, and functional mobility.

As we see with this short history, ethical controversies have attended head transplantation as long as scientists have been doing them. In the past, critiques targeted unwritten social mores and animal rights, and nationalistic ascriptions have tainted scientific advances while perhaps concealing other important ethical assessments. It is possible that because previous BHTs have not necessarily been intended to legitimize human head transplants medically and socially, the ethical challenges have largely been limited to emotional responses, as in Guthrie's case, positioning the research in geopolitical terms, or animal rights protections.

In terms of medical technology, we have been building towards this point for many decades but as often happens, technology has exceeded ethical confirmation. And that is not unusual. The timeline template of an invention is usually an idea, followed by the technology to develop the idea, and then the ethical confirmation or rejection of the whole creation. The innovation to diffusion to acceptance sequence is rarely uniform or directly linear because of how diffusion occurs (forced onto society or voluntarily accepted, for example) and how the creation is socially perceived. These perceptions are typically influenced by whether the invention satisfies a social or personal need or solves a problem, matches socio-cultural tastes, has economic value, elicits minimal fear responses, and conforms to moral sentiments of appropriateness and goodness. Since BHT may be on the horizon, moral analyses have only recently begun.

While at different times in history the heart and even the liver have been endowed with incorporeal qualities such as the center of emotions, we have learned from Hindu cosmology and basic anatomy and physiology that the head carries significant metaphysical, existential, and earthly importance. We also know that the more prominence something holds, the more dangerous or menacing something that threatens it becomes. Perhaps this explains why current critiques of proposed BHT experiments, as we will see, have spawned highly sophisticated and extensive opposition.

Ethical Challenges to Body-Head Transplantation

As we will see in Chapter 2, BHT is a complex surgery with countless potential possibilities for error. Many aspects of the surgery are new and untried. From an immunological perspective, for instance, no tissue as large as a human head has ever been transplanted onto a body. Calculating the immunosuppression dosage and mixture of drugs for this tissue mass is an unprecedented task. Looking further at the surgical process, no fully severed spinal cord has ever been restored, much less transplanted. In fact, the research on repairing spinal cord injuries is in its infancy: to date, no spinal cord tear, partial or full, has ever fully healed. A body has never been decapitated for medical reasons. And while stopping the heart is a common and relatively safe practice in heart surgery, a heart has never been separated from its brain. The existential dilemmas on identity and selfhood that a surviving patient will surely encounter are not only unpredictable but cannot be researched. There are no laboratory trials to gather data on how people will reconcile the phenomenological disorientation they may find themselves with a new body.

The outcomes of a BHT, at least with today's technology, as most have stated, are not likely to favor patients. Estimates foresee three possible medical results. Those who argue in favor of the model to reattach the spinal cords predict a complete recovery. Others are less optimistic. Without a proven method to reattach two disparate ends of spinal cords, a second outcome is that the patient will live but in a state of tetraplegia, which will carry the physical burdens of paralysis and the attending psychological load of adjusting to a new body whose legs or arms cannot move. A third possibility, and the most troublesome one, is the high probability that the patient will not survive and die during or soon after the surgery is completed.

BHT is at once a bioethicist's dream problem and nightmare scenario. As an intellectual exercise, BHT poses a long list of questions about the correctness and viability of its method and the range of outcomes the surgery may cause. Questions as simple as "will it work?" and as metaphysically complicated as "who will the resulting person be?" abound and call for complex philosophical and pragmatic responses. Many bioethicists fear that the procedure lacks moral fortitude because the risks are extraordinarily high and the technology (to date) is untried, making informed consent questionable. The criticisms of BHT are extensive and cover the gamut of human social organization and include questions about BHT's medical rectitude, legal standing, and its psychological, sociological, and philosophical soundness. The critiques are concurrently practical and esoteric, tangible and incorporeal, and general and particular.

Much has been written to address these questions, but these positions have not been integrated into a theoretical whole, and readers are forced to piece their conclusions together to reach a determination of the moral merits of BHT. The intent of this book, therefore, is to assimilate this research into a conceptual totality to help inform IRBs, the medical institution, and the general public in their decisions to accept or reject BHT. To do this, a method and set of axiomatic guidelines are required for assessing the moral turpitude of body to head transplantation. An approach based on social pragmatism will serve this purpose.

Social Pragmatism and Evaluating Body to Head Transplantation

Inspiration, no matter how grand, is subject to limitations. For one, technical obstacles may prevent an idea from coming to fruition. The development of new ideas is confined to the material parameters of mechanical know-how and the availability of resources necessary to make it happen. But even when the expertise and material requisites are present and can turn *le bonne idée* into a tangible reality, social norms and values may create restrictive barriers that hinder its development and dispersion. Innovation and adoption are social phenomena subject to the rules of the social order, the support of those with authority and influence over resources and public sentiment, and the support of the public in general. If the product of the idea is not wanted, innovation can be rejected by the market of public opinion. This is the nature of the clichéd yet imperative proposition: just because something can be done does not mean it should be done.

Time and social changes often force us to renegotiate the "should question". Murray's first successful kidney transplant, for example, was subjected to this criticism as he was accused of "playing God". Now renal transplants are daily fixtures of modern medicine. Heart transplants, which were initially viewed by many as unnatural extensions of life, are now similarly routine. Renal and heart transplants no longer carry ethical baggage that questions their legitimacy.

On some occasions, on the other hand, some ideas fail to overcome moral objections and are rejected on evolving ethical grounds. For example, generally accepted codes of ethics, particularly the Belmont Report of 1979, forbid research on individuals and groups who because of their economic, social, political, or psychological disadvantages can be easily manipulated and exploited by researchers to carry a disproportionate risk burden. Because of these groups' vulnerabilities, researchers, via their professional organizations, and governments have initiated strict rules about protecting certain groups from undue or unequal treatments as research subjects. Researchers must defend including pregnant women, children, and the poor in their investigations by demonstrating how their proposed studies promote and meet expectations of social justice and equity.

The strength of this moral directive has led some publishers and journals to review and investigate the ethical standing of manuscripts they receive. For example, two science publishers, Springer Nature and Wiley, have examined recent research conducted in Xinjiang, China to determine if subjects had given their consent in light of the potent and intrusive surveillance systems employed in that province as a means of social control. Xinjiang is a region that has witnessed mass detentions and alleged severe human-rights violations. Articles written by Chinese police and military officers and legal scholars have been questioned because of their use of population genetics and facial recognition data without consent. These data may have been collected to control and oppress Muslim, Tibetan, and Uyghur minorities (Van Noorden and Castelvecchi 2019).

Ethics are well-founded values of right and wrong that govern what people and social organizations should do. Ethics constitute the interactive synthesis of many social forces. Religious tenets, political opinions, the law, professional imperatives, and culturally normative expectations of behavior are constantly intersecting to produce ethical conduct in medicine and research. Ethics, therefore, are neither universal nor static, and like most norms and values in a society, ethics are relative to time and place. Much of what is unethical today, was normative practice in years past, as witnessed by the current restrictions placed on conducting research on prison inmates. And practices that are ethical in one culture, may be forbidden in another. Embryonic stem cell research is one such medical field that is allowed in some societies but restricted in others because of interpretations of and challenges to its moral foundation.

Ethics, therefore, are subjective. They evolve from social forces that act upon current standards via persuasion, coercion, and social necessity. Furthermore, ethics in general and bioethics in particular are not absolute determinations of correctness; they are subject to time, place, and cultural evolution. In some contexts, such as a scholarly discipline or the law, ethics are usually fairly explicit and often codified elements of a field's normative behavioral character. That is, behavior within the discipline is regulated by ethics. Ethics, however, are also influenced by behavior. In the case of transplantation, once solid organs and faces began to be transplanted without catastrophic outcomes, fears were relaxed and public and professional opinion, and hence formalized bioethics, changed to accommodate the new surgeries. Transplants are no longer questionable ethically though under some circumstances, they are illegal. For example, while kidney transplantation is accepted in virtually all cultures for managing end-stage renal disease, purchasing a kidney is uncompromisingly unethical, except in Iran, a country that allows individuals to sell their kidneys. In the United States, and probably in most places, selling organs was made unlawful to prevent rich and powerful individuals from gaining unfair advantages in accessing health care or coercing underprivileged individuals and families to risk their health by selling their organs to secure basic needs such as food and shelter.

Bioethics, therefore, may best be understood dialectically. They are a result of constant tension between social forces of power, economics, and influence set within a socio-cultural milieux of both sentiments of right and wrong and technology. Each of these dynamics affects the others with the product being the current state of what is and what is not considered ethical. A new technology, in this case body to head transplantation, tests the current state of ethical understanding as a dialectical "antithesis". The degree to which BHT alters the outcome, the dialectical "synthesis", is yet to be known.

Pragmatic Bioethics

The goals of bioethics are to articulate a standard of what is right and what is wrong and enforce those socially defined limits on scientific and medical technology and practice. As such, bioethics is less a method of critical cultural oversight and more of a terrain for ethics experts to review and manage conflicts and problems that arise with emerging technologies. Once new techniques have been tested and produced positive outcomes, and therefore legitimacy, as noted earlier in organ transplants and CTAs, they tend to become value-neutral regardless of the disharmony they created at their onset (Stevens 2000).

The primary method in which bioethicists assess the moral "ripples" caused by new medical technologies is known as principlism, a process in which new technologies are evaluated against a set of established and usually codified moral precepts.

Principlism

Following the atrocities that were conducted in the name of science by Nazi German physicians and scientists and the faux anthropological nature of the Nazi research revealed its insidious political underpinning, governments sought to regulate the behavior of their scholars and medical institutions to protect human research subjects and patients. Regulators were called upon to generate a set of principles to govern scientific behavior and define appropriate research protocols. Over time, a series of ethical rules were implemented globally to assure the rights of research subjects and the responsibilities of scientists. The Nuremburg Codes and the Belmont Report were two of the most important documents that articulated the principles that would serve as the guidelines researchers were obligated to follow. Principlism as the primary mechanism to assess professional conduct was formalized soon thereafter and established the criteria by which moral conflicts are resolved.

Principlism, therefore, is a method of bioethical regulation in which legal and professional entities govern the activities of their constituents. As the name suggests, this paradigm creates a set of universal precepts that apply equally in all situations. To conduct research, investigators must show that their protocols conform to those tenets. Suppositionally, principlism is a hegemonic set of values and an analytical framework that assumes that those tenets are shared by all regardless of culture and social background and that they are independent of cultural and social context.

Despite its humanistic appeal and idealistic intentions, principlism is not a perfect system for evaluating current or proposed medical interventions or research proposals. The presumed universality of the values inherent in this discourse is subject to criticism. For example, one of the most cherished values promoted by principlism is individual autonomy. The right of autonomy insists that each individual is self-determined; that is, people have the right to control their experiences and are capable of making decisions about their own self-interests. Undergoing a medical procedure or participating in an experiment is a voluntaristic choice made only by the patient or study participant and not a physician, investigator, or organization.

Derived directly from the experimental conditions under which an untold number of disenfranchised people suffered unthinkable horrors during the Nazi regime, autonomy is the most powerful principle of bioethics (Wolpe 1998). Self-determination is rooted in a culture of individualism, however, and is not a value found in non-Western and pre-industrial cultures that are collective in nature. In these cultures, decisions are made collaboratively with family members and elders. Decisions are made by social units and consider the needs of the collective as well as that of the individual about whom the decision is being made. Applying Western-type individualism to collectively oriented people by assigning decision-making to individuals and not families may increase uncertainty and cognitive confusion for patients grounded in collectivism. In New Zealand, for instance, the indigenous people, the Māori, have a collective culture in which the family, not the individual, is the basic unit of identity and social action. People think of themselves collectively as members of a family or clan. Because tribal loyalties exceed individual needs in importance and social significance, big decisions such as receiving specialized medical care involve family input and considerations. Western-based health systems, however, fail to consider this aspect of Māori culture. Consequently, Western medical professionals and Māori patients often have difficulty communicating with each other during consultations. With the family as the center of social organization, Māori individuals tend to be deferent to authority and react to medical authorities as if they were family elders. They ask fewer questions about their treatments rather than being proactive in and responsible for their own health behavior, which are expectations among Westerners. Their preference to deliberate options and seek the opinion of others before making up their minds about important decisions often frustrates Westerners who believe the input of family members is intrusive and violates the patient's autonomy and right to self-determination (Sachdev 2001).

Principlism falls short in providing a template for assessing all interventions and innovations because it does not allow for social context and fails to recognize the limits of its universality. Patients left alone to decide to undergo a medical procedure or participate in an experiment may feel abandoned by their physician (Corrigan 2021), confused, and uncertain as to the best course of action. The principle of autonomy allows professionals space to be, in a sense, "off the hook" for making decisions and ignores social and psychological factors that have importance to patients and subjects.

Naturalism

A correlate to principlism is the notion that certain medical interventions should not be promoted and diffused if they are believed to disrupt the natural order of life and create unnatural histories that would not have happened otherwise. This is the notion of "playing God", a charge leveled at Dr. Murray after his historic kidney transplant. In this view, human intervention has its boundaries, and some science technologies cross the line into domains best left to the metaphysical. For theists, this may be God, and for others fate may be the source of otherworldly action. The idea is that certain technologies intrude into a "domain in which another agent is thought to be responsible" (Waytz and Young 2019:8). Humans, in short, should not challenge nature or disrupt the natural flow of what is destined or the will of God.

The norm of maintaining perceived natural process, though, is value-laden "cherry-picking". Not all medical interventions are accused of violating God's will or the natural order of things. Someone who may say that an organ transplant violates nature, may take an antibiotic to cure an infection or have an appendectomy to save their lives from peritonitis. People who invoke this naturalness bias usually do so selectively and employ it when a new medical innovation is perceived as threatening or as challenging to their worldview. The "playing God" argument is a defensive reaction to a technology intuited as dangerous to that which is important in life and the very complexion of how the world works. It is an overly conservative position for assessing innovation and could be considered anti-intellectualism.

Pragmatism and the Study of Potential Consequences

While principlism and naturalness bias provide a moral compass that underscores axiomatic truths about our vision of scholarly behavior, they are not complete methods for measuring innovations. These frameworks, as important as they are, have been challenged on two fronts. One criticism is that they are ethnocentric in that principlism as it is currently presented assumes a Western bias (Kleinman 1997). Second, other factors that may influence decision-making are marginalized. The effects of economic and social marginalization, educational differences, and the lack of professional power play a role in how individuals make decisions (DeVries and Conrad 1998). An example of the latter point is the general distrust of the medical institution that has evolved among African Americans in the United States after decades of differential and inadequate care in comparison to European Americans. A case in point is the lack of trust African Americans place in genetic medicine after centuries of being falsely cast as genetic inferiors and as something less than fully human (Furr 2002).

Evaluating the ethical merits of a technological innovation should be a systematic process that can produce a determination of the innovation's moral value while considering the social context of the innovation and extend beyond the surface level of immediate emotional reactions. A rational approach to the bioethical assessment of a medical practice further seeks to assess all the likely outcomes and alternatives and their relative merits. A socio-economic cost-benefit analysis is certainly part of a rational analysis, but pragmatism also considers the social and psychological implications that an innovation might produce. A pragmatic method strives to mediate the intersection that exists between external (socio-cultural contexts) and internal (needs of patients and doctors) environments (Shilling 2008).

A pragmatic approach to bioethics is particularly concerned with the moral consequences of individual and institutional behavior. Rather than being a theory or set of moral precepts, pragmatism is more of a method for analyzing problems and moral dilemmas brought on by social change. It assumes that reality is not static, a necessary condition for principlism, but dynamic and ever-changing and that the analysis of social change requires evaluating individuals' and groups' interests and intentions. In sum, pragmatism is contextual and emphasizes consequences of a given action. Rather than the ephemeral, pragmatic bioethics centers on the material realities that may result from real or proposed change.

Sociology is an important intellectual force behind pragmatism and impacts bioethics differently than other disciplines, particularly philosophy. Stephen Humphries (2008) contends that philosophers maintain that what should happen cannot be derived from what is happening. We cannot say what is right merely from seeing how things are. In other words, observing what occurs does not proffer sufficient evidence for concluding what ought to be. Following that line of thought, principlism appears overly simplistic and somewhat patriarchal in that the rules of expected behavior are created by a formal mechanism that lacks transparency or consideration of other positions. What should be is a set of value judgements based on perception and subjective interpretation of what "is".

Sociologists, states Humphries, approach bioethics differently and are determined to show how things are and then moving to conclude that things are not as they ought to be. By empirically pointing out problems and their sources, we are better situated to see what does not work and what should be renounced or avoided. Zygmunt Bauman (1993) believes that it is the ambiguity of reality that confounds abstract ethical principles. Moral decisions, therefore, are often ambivalent and lack the ability to adjust to the dynamics of social change.

Sociology provides several conceptual schemes for evaluating the consequences of medical innovation, but for our purposes here, we will rely on an age-old framework to assess the ethical standing of BHT. In 1936, Robert Merton published an important article that articulated a system for determining the origins of unintended consequences of social actions. His typology is an important starting point for studying outcomes of social action and serves as a foundation for building future investigations of the ethics of BHT. While the idea of unintended consequences dates to the 17th

Century and John Locke, Merton provided, for the first time, a heuristic device that can be applied to determine the origins of unanticipated outcomes of a social behavior.

Merton's approach assumes that every behavior intends to produce an imagined outcome, or what he called manifest functions. The manifest function of transplanting a kidney, for instance, is to improve the health and extend the life of the recipient. Merton also realized that reality is not particularly orderly or linear. So, just as every behavior has a manifest function, a latent or unintended function is just as likely to occur. An action may do what it intends, but it may also produce unexpected problems, and Merton sought to identify their origins. His analytical model identifies five sources of latent functions: ignorance, error, imperious immediacy of interests, basic values, and self-defeating prediction.

First, ignorance can detour an anticipated outcome. Not having the knowledge, experience, or skill to produce the desired outcome is likely to produce an undesired result. Ignorance is not necessarily malicious or due to laziness in learning all that is needed to achieve a desired end. Ignorance may take the form of not having the data or information needed to achieve a goal because that knowledge does not yet exist. We do not know what we do not know, as the saying goes, and sometimes we learn that too late.

Concerning BHT, critics have suggested that we do not have the knowledge to mend or attach spinal cords endings. The plan presented to accomplish this, critics charge, is neither adequately tested nor proven to the degree that informed consent can be safely given.

Errors can produce latent consequences. Poor analysis or interpretation of a problem can lead to latent outcomes. Errors may be intentional or unintentional or even random. A slip of the hand in surgery or a discovery of a previously hidden and undetected health condition during a treatment or misreading a probability can lead to unanticipated outcomes.

The BHT process is rife with potential errors. Timing errors, for one, are critical in oxygenating the brain and connecting the spinal cords. Proponents of BHT say that a medically induced coma that would last three to four weeks carries risk that may be unseen. Typically, induced comas last only a few days to two weeks. A month-long coma, though not unprecedented, may present unexpected problems. That physicians contemplating BHT have, as we will see, summarily downplayed the impact of the surgery on the psychological constitution of the body-recipient may be considered an error of exclusion.

Merton's third source of latent functions is imperious immediacy of interests. In this situation, the heightened desirability of social actors to achieve their expected outcomes motivates them to ignore unintentional effects of their behavior. They select their interests over the problems their behavior may or has produced in a gesture of willful ignorance. Sometimes called "short-termism", this is social action that focuses on short-term goals and ignores the long-term. One example of short-termism is the quick approval given to the drug aducanumab, which was designed to treat Alzheimer's Disease. Despite evidence that amyloid-targeting drugs fail to improve the brain health of patients with Alzheimer's, the FDA approved aducanumab. Critics claimed that the FDA's unusual action valued the company's short-term profitability over the public's health and that health care costs could rise. Plus, because the company failed to show that potential benefits outweighed the risk of side effects, researchers were left confused and distrustful of the review process (Servick 2021).

This part of Merton's typology generates discomfort because it leads us to ask for the true motives for performing a BHT if it is not on medical or ethical solid footing. Nonetheless, these questions have been raised. The protocol to conduct BHT has been rejected in Europe and the United States, but has been given life in China, a country that has a history of ethical discordance with Western standards and expectations. Are there underlying motives for conducting the research and doing the surgery there? Sociologists often are curious as to who benefits from a particular social action, and that question is apropos with BHT. We must ask who, besides the patient whose life would be saved, stands to gain from this innovation? Again, these are uncomfortable questions, but some writers have asked them and provided provocative evidence to ground their position. These questions will be considered here in later chapters.

The fourth possible source of unintended consequences are basic values. In this case, the core values underlying an action can produce outcomes that are different or even in contrast to the values. The value of autonomy in medical decision-making is one such example. Our belief in the rights of the individuals to decide their own fates is so strong that we often neglect the social context in which decisions are really made. An unintended consequence of autonomy is disharmony between clinician and patient.

Of course, saving human life is the most cherished value in medicine, but at what cost? We will explore two general points in this area. From a cost-benefit standpoint, is BHT worthwhile to pursue? If there is uncertainty about the possibility that BHT will work, can informed consent be truly given?

Lastly, Merton termed his final source of latent functions self-defeating predictions. This concept describes the case when public prediction of a social development proves false because the prediction itself changed the course of events. The unintended consequence is that the fear of the consequences of a social action propels efforts to find solutions before the problem occurs. One example is that dread of the impact of over-population in the 20th Century led to agricultural enhancements that increased yields that could feed extremely large numbers of people. Oddly, one of those agricultural innovations, the so-called Green Revolution that promoted the use of potent chemical pesticides and fertilizers, had the unintended consequence of increasing cancer rates in rural areas such as Punjab in India. Doom and gloom scenarios often give way to innovations that prevent the disaster from happening.

BHT poses questions here as well. One concerns regulation of surgeries. Surgeries, unlike drugs, are not monitored by regulatory commissions. If the surgery is classified as an experiment, it requires approval by an ethics board. But what if the experiment is deemed a surgery? Then it can proceed without oversight. If the consequences of BHT prove too foreign to established norms and values, then it is possible that innovations in surgery might become subject to ethical review and proof of efficacy before the initial attempt. This is a topic that will be explored in later chapters.

BHT has created a tension within the existing tenets of principlism. The introduction of this radically new procedure challenges our current perceptions of ethical behavior because the range of outcomes is largely unexplored. Our focus here is to concentrate on the likelihood of outcomes. The expressed and objective outcome of BHT is to save the life of a dying individual by transplanting that person's head onto a healthy body. Two additional manifest functions have also been introduced into the fray. Canavero (2022) has stated that BHT will do more than extend the life of the terminally ill, it will also enable human immortality and the permanent settlement of Mars. Our analysis will focus on the first two of these goals, BHT as a life-saving surgery and as the means of achieving immortality. The question is not complicated: What are the likely outcomes of BHT? The answer, however, is convoluted, because BHT entails more than a mechanical and linear cookbook recipe.

A body to head transplant extends into ethical, legal, and psychosocial uncharted waters. Even if the recipe is correct and a head can be reattached to a body, there are implications and potentially negative consequences that have not yet fully coalesced into a position of determining a decree to allow or prohibit the experiment. If critics are to be believed, there remains both a material and metaphysical fissure between the imagination that created BHT and the reality of its outcomes. In BHT, imagination and reality are at odds and the gap between them constitutes a test of our morals.

References

Alberti, F.B. and V. Hoyle. 2021. Face transplants: An international history. Journal of the History of Medicine and Allied Sciences, 76(3): 319–345. doi.org/10.1093/jhmas/jrab019.

Bain, J.R. 2000. Peripheral nerve and neuromuscular allotransplantation: Current status. Microsurgery, 20(8): 384–388. doi.org/10.1002/1098-2752(2000)20:8<384::AID-MICR7>3.0.CO;2-W.

Barker, J.H., A. Furr, M. Cunningham, F. Grossi, D. Vasilic, B. Storey et al. 2006. Investigation of risk acceptance in facial transplantation. Plastic and Reconstructive Surgery, 118(3): 663–670. doi.org/10.1097/01. prs.0000233202.98336.8c.

Bauman, Z. 1993. Postmodern Ethics. Oxford: Blackwell, 1993. Blackwell, Oxford.

Bennett, C. 21 August 1995. Letter to the editor: Cruel and unneeded. New York Times.

Bhattacharya, S. 23 July 2003. First human tongue transplant successful. New Scientist. newscientist.com/article/dn3964-first-human-tongue-transplant-successful/ (accessed 3 January 2023).

Brouha, P., D. Naidu, M. Cunningham, A. Furr, R. Majzoub Grossi et al. 2006. Risk acceptance in composite-tissue allotransplantation reconstructive procedures. Microsurgery, 26(3): 144–149. doi.org/10.1002/micr.20227.

Brown, G.W. 1921. The Human Body in the Upanishads. Facsimile Publisher, New Delhi.

Canavero, S. 10 April 2022a. World's 1st Human Head Transplantation—Dr Sergio Canavero—Neurosurgeon [YouTube Channel]. youtube.com/watch?v=KY_rtubs6Lc.

Corrigan, O. 2003. Empty ethics: The problem with informed consent. Sociology of Health and Illness, 25(7): 768–792. doi.org/10.1046/j.1467-9566.2003.00369.x.

Daniel, R.K., E.P. Egerszegi, D.D. Samulack, S.E. Skanes, R.W. Dykes and W.R.J. Rennie. 1986. Tissue transplants in primates for upper extremity reconstruction: A preliminary report. The Journal of Hand Surgery, 11(1): 1–8. doi.org/10.1016/S0363-5023(86)80091-0.

DeVries, R. and P. Conrad. 1998. Why bioethics needs sociology. pp. 233–257. *In*: R. DeVries and J. Subedi [eds.]. Bioethics and Society: Constructing the Ethical Enterprise. Prentice Hall, Upper Saddle River, NJ.

Diep, G.K., Z.P. Berman, A.R. Alfonso, E.P. Ramly, D. Boczar, J. Trilles et al. 2021. The 2020 facial transplantation update: A 15-year compendium. Plastic and Reconstructive Surgery - Global Open, 9(5): e3586. doi.org/10.1097/GOX.0000000000003586.

Dubernard, J.M., E. Owen, G. Herzberg, M. Lanzetta, X. Martin, H. Kapila et al. 1999. Human hand allograft: Report on first 6 months. The Lancet, 353(9161): 1315–1320. doi.org/10.1016/S0140-6736(99)02062-0.

Duran, X. 2018. Els trasplantaments ala literatura: Un empelt de ciència ificció. Mètode Revista de Difusió de La Investigació, 8. doi.org/10.7203/metode.8.10498.

Elliot, D. 2 July 2013. Human head transplant is "bad science," says neuroscientist. CBS News. cbsnews.com/news/human-head-transplant-is-bad-science-says-neuroscientist/ (accessed 8 December 2022).

Emmerich, N. 2021. Ethos and eidos as field level concepts for the sociology of morality and the anthropology of ethics: Towards a social theory of applied ethics. Human Studies, 44(3): 373–395. doi.org/10.1007/s10746-021-09579-2.

Errico, M., N.H. Metcalfe and A. Platt. 2012. History and ethics of hand transplants. JRSM Short Reports, 3(10): 1–6. doi.org/10.1258/shorts.2012.011178.

Fageeh, W., H. Raffa, H. Jabbad and A. Marzouki. 2002. Transplantation of the human uterus. International Journal of Gynecology & Obstetrics, 76(3): 245–251.doi.org/10.1016/S0020-7292(01)00597-5.

Francois, C.G., W.C. Breidenbach, C. Maldonado, T.P. Kakoulidis, A. Hodges, J.M. Dubernard et al. 2000. Hand transplantation: Comparisons and observations of the first four clinical cases. Microsurgery, 20(8): 360–371. doi. org/10.1002/1098-2752(2000)20:8<360::AID-MICR4>3.0.CO;2-E

Furr, A. 2022. The Sociology of Mental Health and Illness. Sage, Thousand Oaks, CA.

Furr, A., M.A. Hardy, J.P. Barret and J.H. Barker. 2017. Surgical, ethical, and psychosocial considerations in human head transplantation. International Journal of Surgery, 41: 190–195. doi.org/10.1016/j.ijsu.2017.01.077.

Furr, L.A. 2014. Facial disfigurement stigma: A study of victims of domestic assaults with fire in India. Violence Against Women, 20(7): 783–798. doi. org/10.1177/1077801214543384.

Goldman, M. 1987. Lister Ward. Adam Hilger, Bristol.

Goodrich, M.J. 2020. Medical violence and medieval miracle of the black leg. Synapsis. medicalhealthhumanities.com/?s=goodrich (accessed 12 November 2022).

Guimberteau, J.C., J. Baudet, B. Panconi, R. Boileau and L. Potaux. 1992. Human allotransplant of a digital flexion system vascularized on the ulnar pedicle: a preliminary report and 1-Year follow-up of two cases. Plastic and Reconstructive Surgery, 89(6): 1135–1147.

Guthrie, C.C. 1912. Blood-Vessel Surgery and Its Applications. Longmans, Green. New York.

Harikrishnan, P. 2019. Head transplant: Surprising and interesting facts from the Hindu mythology. Journal of Craniofacial Surgery, 30(5): 1329–1330. doi. org/10.1097/SCS.0000000000005125.

Hirabayashi, S., H. Kiyonori, A. Sakurai, K.T. Edson and F. Osamu. 1988. An experimental study of craniofacial growth in a heterotopic rat head transplant. Plastic and Reconstructive Surgery, 82(2): 236–243.

Hoffenberg, R. 2001. Christiaan Barnard: His first transplants and their impact on concepts of death. BMJ, 323(7327): 1478–1480. doi.org/10.1136/bmj.323.7327.1478.

Hovius, S.E.R., H.P.J.D. Stevens, P. van Nierop, W. Ratng, R.M. van Strik and J.C. van der Meulen. 1992. Allogeneic transplantation of the radial side of the hand in the rhesus monkey. Plastic and Reconstructive Surgery, 89(4): 700–709.

Hughes, M.J. 1998. The Social Consequences of Facial Disfigurement. Routledge, London.

Humphreys, S.J. 2008. The sociology of bioethics: The 'is' and the 'ought.' Research Ethics Review, 4: 47–51.

Jones, J.W., S.A. Gruber, J.H. Barker and W.C. Breidenbach. 2000. Successful hand transplantation—One-year follow-up. New England Journal of Medicine, 343(7): 468–473. doi.org/10.1056/NEJM200008173430704.

Jones, J.W., E.T. Üstüner, M. Zdichavsky, J. Edelstein, X.P. Ren, C. Maldonado et al. 1999. Long-term survival of an extremity composite tissue allograft with FK506–mycophenolate mofetil therapy. Surgery, 126(2): 384–388. org/10.1016/S0039-6060(99)70181-9.

Jones, T.R., P.A. Humphrey and D.C. Brennan. 1998. Transplantation of vascularized allogeneic skeletal muscle for scalp reconstruction in a renal transplant patient. Transplantation, 65(12): 1605–1610.

Jović, N.J. and M. Theologou. 2015. The miracle of the black leg: Eastern neglect of Western addition to the hagiography of Saints Cosmas and Damian. Acta Medico-Historica Adriatica: AMHA, 13(2): 329–344.

Kleinman, A. 1997. Writing at the Margin: Discourse between Anthropology and Medicine. University of California Press, Berkeley.

Konstantinov, I.E. 2009. At the cutting edge of the impossible: A tribute to Vladimir P. Demikhov. Texas Heart Institute Journal, 36(5): 453–458.

Levi, D.M., A.G. Tzakis, T. Kato, J. Madariaga, N.K. Mittal, J. Nery et al. 2003. Transplantation of the abdominal wall. The Lancet, 361(9376): 2173–2176. doi. org/10.1016/S0140-6736(03)13769-5.

Mackinnon, S.E. 1996. Nerve allotransplantation following severe tibial nerve injury: Case report. Journal of Neurosurgery, 84(4): 671–676. doi.org/10.3171/jns.1996.84.4.0671.

Mackinnon, S.E., V.B. Doolabh, C.B. Novak and E.P. Trulock. 2001. Clinical outcome following nerve allograft transplantation. Plastic and Reconstructive Surgery, 107(6): 1419–1429.

McCrone, J. 2003. Monkey business. The Lancet Neurology, 2(12): 772. doi. org/10.1016/S1474-4422(03)00596-9.

Niu, A., K. Shimazaki, Y. Sugawara, T. Mizui and N. Kawai. 2002. Heterotopic graft of infant rat brain as an ischemic model for prolonged whole-brain ischemia. Neuroscience Letters, 325(1): 37–41. doi.org/10.1016/S0304-3940(02)00213-6.

Nizzi, M.C., S. Tasigiorgos, M. Turk, C. Moroni, E. Bueno and B. Pomahac. 2017. Psychological outcomes in face transplant recipients: A literature review. Current Surgery Reports, 5(10): 26. doi.org/10.1007/s40137-017-0189-y.

O'Carroll, S. February 16, 2013. The history of the two-headed dog experiment. The Journal.Ie. thejournal.ie/two-headed-dogs-794157-Feb2013/ (accessed 4 December 2022).

Pomahac, B., R.M. Gobble and S. Schneeberger. 2014. Facial and hand allotransplantation. Cold Spring Harbor Perspectives in Medicine, 4(3). doi. org/10.1101/cshperspect.a015651.

Reece, E. and R. Ackah. 2019. Hand transplantation: The benefits, risks, outcomes, and future. Texas Heart Institute Journal, 46(1): 63–64. doi.org/10.14503/THIJ-18-6739.

Ren, X., M.V. Shirbacheh, E.T. Ustuner, M. Zdichavsky, J. Edelstein, C. Maldonado et al. 2000. Osteomyocutaneous flap as a preclinical composite tissue allograft: Swine model. Microsurgery, 20(3): 143–149. doi.org/10.1002/(SICI)1098-2752(2000)20:3<143::AID-MICR9>3.0.CO;2-9.

Sachdev, P.S. 2001. The impact of colonialism on the mental health of the New Zealand Maori: A historical and contemporary perspective. pp. 15–45. In: D. Bhugra and R. Littlewood [eds.]. Colonialism and Psychiatry, Oxford University Press, New Delhi.

Schillace, B. 2022. Mr. Humble and Dr. Butcher: A monkey's Head, the Pope's Neuroscientist, and the Quest to Transplant the Soul. Simon & Schuster, New York.

Schuind, F., C. Van Holder, D. Mouraux, C. Robert, A. Meyer, P. Salvia et al. 2006. The first Belgian hand transplantation—37 month term results. Journal of Hand Surgery, 31(4): 371–376. doi.org/10.1016/J.JHSB.2006.01.003.

Servick, K. 2021. Alzheimer's drug approved despite doubts about effectiveness. Science. doi.org/10.1126/science.abj8372 (accessed 15 November 2022).

Shilling, C. 2008. Changing Bodies: Habit, Crisis and Creativity. Sage, Los Angeles.

Shirbacheh, M.V., J.W. Jones, M. Zdichavsky, W.C. Breidenbach and J.H. Barker. 1998. The feasibility of human hand transplantation. Seventh IFSSH Congress, 24–28, Vancouver.

Stark, G.B., W.M. Swartz, K. Narayanan and A.R. Møller. 1987. Hand transplantation in baboons. Transplantation Proceedings, 19(5): 3968–3971.

Stephenson, H.E., R.S. Kimpton and G.M. Masters. 2001. America's First Nobel Prize in Medicine or Physiology: The Story of Guthrie and Carrel. Midwestern Vascular Surgery Society and Hugh E. Stephenson, Boston.

Stevens, T. 2000. Bioethics in America: Origins and Cultural Politics. Johns Hopkins University Press, Baltimore.

Strome, M., J. Stein, R. Esclamado, D. Hicks, R.R. Lorenz, W. Braun et al. 2001. Laryngeal transplantation and 40-month follow-up. New England Journal of Medicine, 344(22): 1676–1679. doi.org/10.1056/NEJM200105313442204.

Sugawara, Y., S. Hirabayashi and K. Harii. 1999. Craniofacial growth in a whole rat head transplant: How does a non-functional head grow? Journal of Craniofacial Genetics and Developmental Biology, 19(2): 102–108.

U.S. Department of Health and Human Services. 2023. Organ Procurement and Transportation Network. optn.transplant.hrsa.gov/news/2022-organ-transplants-again-set-annual-records-organ-donation-from-deceased-donors-continues-12-year-record-setting-trend/.

Ustuner, E.T., R.K. Majzoub, X.P. Ren, J. Edelstein, C. Maldonado, G. Perez-Abadia et al. 2000. Swine composite tissue allotransplant model for preclinical hand transplant studies. Microsurgery, 20(8): 400–406. doi.org/10.1002/1098-2752(2000)20:8<400::AID-MICR10>3.0.CO;2-Z.

Van Noorden, R. and D. Castelvecchi. 2019. Science publishers review ethics of research on Chinese minority groups. Nature, 576(7786): 192–193. doi.org/10.1038/d41586-019-03775-y.

Waytz, A. and L. Young. 2019. Aversion to playing God and moral condemnation of technology and science. Philosophical Transactions of the Royal Society B: Biological Sciences, 374(1771). doi.org/10.1098/rstb.2018.0041.

White, R.J., L.R. Wolin, L.C. Massopust, N. Taslitz and J. Verdura. 1971. Primate cephalic transplantation: Neurogenic separation, vascular association. Transplantation Proceedings, 3(1): 602–604..

Wolpe, P.R. The triumph of autonomy in American bioethics: a sociological view." Bioethics and society: Constructing the ethical enterprise (1998): 38–59.

World Health Organization. 2020. International Report on Organ Donation and Transplantation Activities. transplant-observatory.org (accessed 9 November 2022).

2

What Might Happen?
The Benefits and Risks of
Body-Head Transplantation

Chapter Summary

If, as discussed in Chapter 1, ethical systems are designed primarily to assess innovations, then we should consider in detail those innovations that might challenge current ethical norms and values. The bioethical implications of this experiment are centered around the possible consequences of those innovations. The question that guides this chapter asks if the risks of the introduction of the new surgical theory proposed in the BHT model are within limits of social and professional mores.

This chapter describes the general surgical plan for conducting a BHT as proposed by Sergio Canavero. Not every medical detail is included as some are rather common procedures that spark no ethical worries in and of themselves. Instead, I describe the main and novel steps outlined in Canavero's agenda. Following this review, the chapter highlights the points of primary risk. Many of these risks, such as intense central pain and paraplegia, are more openly discussed in the literature. Others, such as memory loss and psychiatric distress, however, are more of the latent or unintended variety and quite discerning.

One of the core tenets of principlism is the responsibility of physicians to communicate the risks of the therapies and interventions they extend to patients. Patients then assess that information to determine if the advantages of the proffered medical treatment outweigh the potential costs. For patients to be truly self-determined, that is to make decisions that are in their own best interests, physicians are ethically obligated to inform patients in lay language of the potential outcomes of a procedure based on research and their own experiences in administering that treatment.

*   *   *

Most medical interventions patients encounter are tried and true, making the moral prerogative to inform patients of their risks a relatively routine and mundane step in acquiring informed consent. For new or experimental interventions, however, which are unfolding at lightning speed, data are often lacking, and communicating risk has become complicated, often requiring patients to hold some degree of scientific and medical knowledge to completely comprehend the intended goals of the treatment in proportion to side effects or failures. Adding to the complexities of modern medicine and their attendant complications in communicating risk is that socio-cultural perceptions of risk have changed since the core tenets of principlism were first proposed.

In health and medicine, risk was initially a quantified and neutral term (Lupton 1993). In its quantified form, risk was equated with the calculation of the statistical probability of an outcome. Presenting the odds of success or failure was considered satisfactory and sufficient information for physicians and patients to assess a proposed treatment plan. As a neutral term, risk was devoid of meaning beyond the mathematical. Risk was cut-and-dried; it was simply a number, a chance, based on past experience.

More recently the meaning of risk has morphed into a more complex idea. Rather than a neutral term of statistical chance, morals and values are now attached to the numbers so that risk has become as much an emotional concept as a mathematical one. Risk, as Mary Douglas (1990) observed, is now equated with danger. Nowadays risk means trouble, and the mere mention of the word evokes emotions of dread and fear. Risks amount to hazards, not odds, that imply and perhaps forebode disastrous ends. Managing risks, which is now a professional occupation, infers regulating the origins of those hazards to reduce liability and harm and raise confidence and trust.

The progression of risk from neutral calculations of consequences to a term laden with emotions mirrors the evolution of the origins of risk encountered by people living in Western societies. As Ulrich Beck (1992) has noted, risk perception and definition spring from the types of problems people now face in daily life. In times past, the risk exposure of most people was confined to the forces of nature. The hazards faced in everyday life had more to do with anxiety about insufficient rainfall to sustain crops, the location and health of animal herds, and attacks by wild animals.

In contemporary life, people are more likely to face dangers that are human-made rather than nature-made. These are known as "manufactured risks"; that is, they are the unsafe and threatening products of human activity. Technology is causing potentially catastrophic risks to the environment and human health and well-being. Lifestyle choices, such as the consumption of drugs that threaten life and psychosocial health, are deeply connected to social organization and individuals' places in society (Furr 2022). As we learned during the Covid-19 pandemic, people believe that science can both create and mitigate risks. Deep-seated distrust of science (paired with the government's willingness to partner with scientists) led to unnecessary contentiousness over the origins of the Covid virus and how to manage the pandemic, assuming of course that people believed the pandemic was real for there were many who did not. The emotional responses to Covid are attributable to the disenchantment people feel because of the dominance of manufactured risks in everyday life. We are reminded daily that foods we like to eat are not healthful, that strangers will kidnap and murder us, that people carry guns to protect themselves from people who carry guns, and that the earth is going to die because of warming due to human activity.

Meanings beyond fate or God's will, the explanations given to explain dangers caused by nature, are now attached to the concept of risk, meanings that are often tainted by political and social values and often without regard to the factual conditions of outcomes. People largely believe that other people pose more risk and danger than does nature, and they expect to be protected from human error, capriciousness, negligence, and greed, perhaps the root causes of most manufactured risks. Risks are often connected to moralistic propositions that may or may not have anything to do with evidence-based science. These values can, and likely do, have more influence in decision-making than actual statistical risk because the idea of neutrality in risk perception no longer really exists. In this mindset, human activity that creates risk has political, social, or economic motives at its underlying core.

Effective risk communication now should extend the conversation beyond simple likelihoods of success or failure or side effects. Now risk must be discussed in the context of patients' values, social position, and science literacy. Understanding and avoiding risk is as influenced by cynicism, distrust, disenchantment, and alienation as it is by medical fact and practice wisdom. Similarly, acceptance of certain risks may be based on unrelated criteria. Perceptions of risk and decision-making can be affected by narcissism ("this will work for me though not for others"), denial ("I don't believe the risk is real"), social pressure ("my social peers tell me what is right and wrong, and I want to fit in"), or desperation ("I have to do this or something bad will happen to me"). If risk is understood as danger, then the cognitions at work are those that individuals use to process danger, not rationally calculating odds and basing a decision on "the numbers".

Good risk communication, therefore, takes into consideration the psychosocial dynamics that influence decision-making centered around accepting or rejecting potential dangers. Since principlism only requires the communication of what might happen and the chances of those outcomes happening, most of the information that people use to give informed consent is untouched in the conversation. Principlism leaves decisions solely in the hands of patients, and patients will employ all avenues of emotional, cognitive, and social input to help make those decisions. Subsequently disconnects between medicine and the public surface and conflicts may even arise. One need not go far to find recent examples: false "truths" about the relationship between vaccines and autism, denial of Covid-19, and the many incorrect, yet believed, preventions and treatments for Covid-19.

Bioethics is the arena in which this discourse plays out in medicine, philosophy, and the social sciences. Pragmatic bioethics is particularly interested in analyzing the social context in which a perception of risk is generated and maintained. In this sense, bioethics must evolve with changing social conditions, reacting to challenges from non-medical sources, and determining how to assess and then adopting or abandoning new technologies. Regarding the latter, bioethics must determine if emerging technologies are consistent with extant social norms and established medical practice.

Associating risk with danger is the new operating paradigm for bioethics. Health research, especially epidemiology, conceptualizes risk as the probability of a dangerous or unhealthful event or occurrence. While risk is still presented as odds, such as the likelihood of smokers getting lung cancer, it is not necessarily presented or heard without a moral compass giving meaning to that number. Lung cancer is sometimes stigmatized because of a lifestyle choice, using tobacco, associated with getting the disease. Similarly, cirrhosis of the liver due to chronic alcohol abuse is characterized as a moral condition by many who also believe that alcoholic liver disease patients should not have priority for liver transplants over needy patients whose livers have failed through no fault of their own.

If risk equals both the probability of a negative outcome and the perception of danger, when actual risk probability is unknown, and therefore subtracted from the equation, then its association with danger is the only definition that remains. Without the mathematical odds of the possible outcomes, we must assume that any potential risk factor is dangerous because of the cultural context in which we now identify and perceive the idea of risk. In these cases, unseen risks are therefore unseen dangers.

When we do not have quantifiable independent variables derived from studies based on a sufficiently large sample, the actual likelihood of an outcome cannot be produced and communicated to patients. Risk, and therefore danger potentials, can, however, be inferred through a deductive method in which possibilities may be predicted through extrapolation from similar situations, theoretical assumptions, and by comparing analogous events. Using these methods to determine risk in experimental medicine, physicians and researchers have a moral obligation to think through all the possibilities of their proposals and map out potentialities of what could go wrong or how the outcomes will change the pre-experimental patient. How those conclusions are interpreted become the perception of risk for new and experimental medical interventions.

Because there are no data on BHT among humans and the animal experiments that have been conducted are considered ungeneralizable to humans, we are forced to use softer, that is, nonstatistical, language to describe what is likely to happen to individuals who undergo a BHT at this time in history. The implications of BHT extend beyond its proposed life-saving outcome that will return a terminally ill person to full health. They also include all its possible consequences. In the absence of substantive data to predict what will happen in this experiment, we have to treat all possible outcomes as likely and even probable, the logic being that any risk equals danger.

Many of the potential risks of BHT have been downplayed or even dismissed by proponents of the BHT surgical model. Advocates of BHT have not addressed adequately or comprehensively identified all the potential dangers of the BHT proposal or concept. In fact, they are on record as summarily scorning their critics both in writing and in public discussions. For example, in a 2016 interview on the television program Good Morning Britain, Sergio Canavero, the leading champion of BHT, responded to host Piers Morgan's questions and the comments from an in-studio consultant called Dr. Hilary with derision by saying "these people [the critics] have no idea what they are talking about." He accused the in-studio consultant and other critics of not having read his papers on BHT (without asking for verification of that claim) and deflected further questioning by saying the nay-sayers should "stay tuned because more news is coming." When asked about the chances that a patient will survive the surgery, Canavero declared a specific number: "90 percent". Through his own admission, this figure was derived from research he had yet to conduct on cadavers and transplanting a head on two brain-dead patients. This rhetorical style is not untypical, which has been commented on by critics such as Paul Wolpe (2017) who have accused Canavero of conducting medical and ethical debates via the media rather than through scholarly journals and thereby skirting medical details. The 90 percent figure is indeed misleading and cannot be studied for verification, but it may influence individuals into believing that BHT is almost assuredly safe. As Wolpe states (2017: 207), "it is precisely this continuous dance around the science and medicine that so concerns those who look seriously at Ren and Canavero's claims".

This chapter will review the basic surgical plan for conducting a human BHT and will discuss, qualitatively, the potential outcomes and their impact on the health and humanity of a patient. The specific question at hand: what are the dangers of body and head transplantation? We are defining the risks of BHT as manufactured risks because the experiment has, as we shall soon see, numerous risks whose resolutions have not been detailed. BHT is a socially constructed process in which natural process contained within the continuity of a whole body are disrupted and replaced. BHT is not a usual transplant. Rather it transforms and reconstructs the human form into an aberration of the natural body. The decision to design, conduct, and subject oneself to BHT is a human decision. Whatever happens to patients afterwards is a product of a human idea and the socially constructed context that legitimated it.

What Could a Patient Expect to Happen?

In 2015, Canavero announced that Valery Spiridonov, a 30 year-old IT specialist and successful entrepreneur, had volunteered to undergo the first body to head transplant operation. As is well documented in the public media, Spiridonov has Type 1 Spinal Muscular Atrophy (SMA 1), a rare muscle-wasting condition also known as Werdnig-Hoffman disease. The symptoms of SMA 1 are quite serious and usually present at birth or soon thereafter. Werdnig-Hoffman causes devastating muscular weakness, poor muscle tone, and little motor development. As the disease progresses with the continued loss of motor neurons in the spinal cord and brainstem, children with SMA 1 experience trouble swallowing, breathing, and holding their heads upright. Individuals with Werdnig-Hoffman possess normal intelligence, but they are unable to sit, stand, or walk (Emmady and Brodle 2022). Because there is no cure for this dreadful disease, Spiridonov presented as an ideal candidate for a BHT. Spiridonov's surgery, it was announced to the world, would be conducted in 2017.

Most of the information known about the relationship between Canavero and Spiridonov is learned through media stories and interviews. According to one report, the two had only met via Skype as the decision was made to accept Spiridonov as a patient (Nelson 2015). They later connected in person, and a few of their meetings were captured on video and posted online. One example is the Good Morning Britain program referenced earlier in this chapter in which Canavero was interviewed remotely while Spiridonov was present in ITV's studio. Canavero and Spiridonov did not often appear together on the internet, though scores of interviews with and stories about them are revealed in a search on youtube.com. There are too many other stories and interviews to count on the internet. For instance, in one of these stories we learned that Spiridonov, though personally financially secure, did not have the means to pay for the expensive experimental surgery. To raise money, one media story reported that Spiridonov ran an online venture selling hats, t-shirts, mugs, and iPhone covers with an image of his head on a new body (Bonvillian 2016).

By 2018, however, Spiridonov had abandoned the surgery and withdrew from participating. Again, per media reports, Spiridonov postponed the operation because he believed the surgeons had lost interest in him since he could not pay for the surgery. He later moved to the United States and focused on his own work on artificial intelligence and robotics, inventing a "smart" wheelchair with voice control (Alekseev.biz 2022). He also had started his own company and married a woman he met at his workplace. He and his wife later had a son who did not inherent Werdnig-Hoffman disease. The Russian national has stated that the surgery would take away too much time from his new family and that his health had stabilized. In his mind, the conditions of his life had changed for the better and the appeal BHT began to wither. He was quoted by the Canadian publication National Post as saying, "I've got my own things to do" (Kirkey 2019). Another article suggested he had lost faith in the surgery (Schillace no date).

It would be sound to assume that any reasonable person would opt for a corrective surgery if the risks were manageable. Each day people throughout the world make the decision to replace an arthritic hip or knee, undergo an emergency appendectomy, or agree to a major life-saving surgery such as a kidney or liver transplant. But experimental surgeries are different. Few want to be first, and that's simple enough to understand: the outcomes are uncertain. Even oft-practiced surgeries with relatively low success rates such as those to remove malignancies are often deemed acceptable by patients who play the odds of a successful surgery against the cancer. But if life is acceptable without the surgery even if the illness persists suggests that living with the illness offers better odds of acceptable quality of life than what the surgery offers. The cost and benefit analysis shifts, and the lived perils are preferred to the unknown, especially when one of those unknowns is whether they will wake up from the anesthesia.

In any case, both surgeon and patient decided to move on.

Head transplantation did not receive ethical or medical approval in Spiridonov's native Russia, Europe, or the United States. The BHT team decided to relocate to China, which provided considerable support for the research and has fewer ethical criteria to satisfy than Russia and the western countries. It first appeared that Spiridonov's operation would have been conducted there, but after he withdrew, Canavero was later quoted in the Canadian press that the Russian was never considered as a candidate for the surgery in that country "for obvious cosmetic reasons" (Kirkey 2019). Several media sites have reported Canavero saying he has a long list of volunteers willing to undergo the untried surgery. Unlike Spiridonov, the names of these potential patients and their health status and locations have been withheld.

Over five years have passed since the announced target date of December 2017, and no BHT is known to have been conducted. Although the surgery has not yet been performed, the plan remains and continues to attract attention from bioethicists. Into the 2020s, Canavero continues to promote the experiment in the media and no physician[1] involved in the development of the surgical model has publicly called for an end to the drive to perform a BHT; therefore, BHT remains "in play" and lingers as a potentiality.

Had Spiridonov remained a viable candidate for a BHT, what could he and the aforementioned unidentified roster of volunteers expect to happen if they underwent this revolutionary and paradigm changing surgery? With the exception of the catastrophic outcome of death, success and failure are not necessarily absolute or categorical states. There is a plethora of possible outcomes beyond the two discrete categories of (a) success as defined as perfect health, and (b) death. In other words, success is likely not without qualifications. Survival may remove the terminal health condition that threatened the patient's life and led to the BHT, but other problems may replace it and make life just as untenable as it was before the surgery. In short, there is every reason to question the posture that BHT offers a complete "cure" where patients wake up from anesthesia, complete a program of physical and occupational therapy, and walk away happy to have a new life free of the terminal disease that had once threatened their very existence, which is what BHT's proponents say will happen.

Before moving on to the multifarious bioethical problems posed by transplanting human heads, we must first have an idea of what the surgery entails and how it might be conducted.

In the most basic of terms, what is head transplantation and how might it work? Is there a disconnect between what is fantasy and what is real? Let's review in lay language how BHT might work and the ethical questions it has generated.[2]

---

[1] Kirkey (2019) reports that Xiao Ping Ren, Canavero's research partner and collaborator, has stated that his, Ren's, true professional goal is to repair spinal cord injuries not to transplant heads, a puzzling statement given the several articles Ren has authored and co-authored endorsing and promoting body-head transplantation.

[2] In addition to the original sources in Canavero (2013; 2015b), Canavero et al. (2016a), and Ren et al. (2019), good medical reviews of the steps of a BHT are found in Gkasdaris and Birbilas (2019) and Barker et al. (2018).

The Blueprint of a Body-Head Transplantation

The surgical requirements to perform a BHT would constitute the most ambitious medical endeavor ever undertaken and command the expertise of many surgical and medical specialists. What differs from the earlier animal head transplants conducted by Robert White is that physicians have now published a detailed plan with intentionality—a stated aim of separating and reattaching a human head onto another body— and dictating the steps to make it possible. This surgical protocol was devised by Italian neurosurgeon Sergio Canavero, who labeled it "head anastomosis venture", or HEAVEN, and elaborated upon by his associate Xiaoping Ren (Canavero 2013; Ren et al. 2019). It is HEAVEN, along with its accompanying research, that will be reviewed and scrutinized for its ethical merits. Whether the design will achieve its goal of complete health for the head donor is open to debate. Doubting it will work, many physicians and bioethicists have called the protocol pseudo-science and fantasy. Nonetheless, HEAVEN exists, and a handful of scholars have devoted time and resources to exploring its possibilities.

Before the Surgery Begins

Before the first scalpel cut is made, a team contemplating a BHT must expend considerable effort and expense on preliminary research and administrative organization. After the proof of concept animal and cadaver research is conducted and the results indicate that the surgery will work thereby making truly informed consent possible, a first step would be to receive a hospital's permission and commitment to provide the surgical and recovery space, staff, and resources. As will be shown in Chapter 3, a hospital hosting the first BHT, if not the first several, would be wise to consult its legal team to map out its institutional liability. Given the experimental nature of the surgery and the range of outcomes that are possible, the supporting hospital could easily find itself a defendant in a lawsuit should outcomes substantially different than a complete cure occur.

    While negotiating with hospitals, the surgical staff would need to procure a two-front financial foundation. First, costs for the hospital and personnel must be covered in advance of the surgery, and second, economic support for patients and their families must be arranged. Although no itemized budget has been released, Canavero told *USA Today* that the first BHT will cost about $100 million (Hjelmagaard 2017). Because there is no detailed or itemized cost estimation, this figure is not particularly reliable. Nonetheless, the actual cost of performing and recovering from a BHT would be extraordinarily high, and a patient of ordinary means could not afford it.

Adding to the cost of the surgery, a surviving patient would be economically dependent for a long period of time and would require income to pay for housing, nutrition, health care, insurance, and entertainment during the long period of recuperation that will follow the surgery. Presumably patients' families would provide much of the aftercare, which means cutting into their ability to work, tapping their savings, and perhaps putting them into debt. In addition to a host of other medications, expensive immunosuppressing drugs would be required for the remainder of patients' lives. Without significant family and external financial support, the economic burdens BHT would impose on patients and their families could easily cause chronic social and psychological stress and hinder recovery.

As capital is raised, the surgical team must seek approval from relevant ethics review panels, or Institutional Review Boards (IRB) as they are known in the United States. The proposal sent to the IRBs would necessitate the description of every step in the process in precise detail, indicating points of risk and likelihood of failure or error. The proposal must also explain how the head donor patient would be recruited and how that person's consent would be secured. It would also describe how the donor body would be located and identified, assessed, and consented. In most countries, legal codes and extant bioethics practices would allow or require families to provide consent for a brain-dead relative's participation. Disposal of the remains, the body donor's head and the head donor's body, would also be specified in the proposal.

The IRB would outline all the steps of recruiting a volunteer head donor as well as the preoperative tests that would be required. In addition to a complete medical exam and determination that patients qualify and can withstand the surgery, psychological testing protocols would be included in the IRB proposal. Having learned from past CTA experiences, head donors would have to undergo an extensive psychological evaluation to ensure that they have the emotional, cognitive, and behavioral strength and presence of mind to:

- cope with the risks the surgery entails
- endure and complete the long, difficult, and likely painful rehabilitation that would follow
- stomach the toxic side effects of the lifelong immunosuppression required to prevent rejection
- comply with post-operative treatments, and
- perhaps most critically, manage the phenomenological disruptions of perceptions of identity and sensory stimulation presumed to follow the acquisition of a totally new body.

The proposal should also state how patients' psychological data would be interpreted to determine psychological fitness.

The IRB proposal would also require surgeons to provide a complete assessment of the physiological and psychological risks as well as evidence-based probabilities of potential outcomes for the patient. Since, at this point, no experimental data from which to draw conclusions of risk is available. Any assessment of risk would be patient specific.

With IRB approval in hand, surgeons could only then begin the process of identifying two individuals willing to participate in the experiment: a body donor and a head donor, the latter of which will survive in the sense that this is the person whose consciousness and sense of identity will, presumably, awaken after being revived from the proposed post-operative coma. The other participant is a brain-dead individual whose body is suitable for transplantation and whose family consents to donate their loved one's body. The body donor's blood type, crossmatch testing, and general health must present compatibility with the head. In addition, appropriate sex, height, skin tone, and the head donor's acceptance of any idiosyncratic markings on the body such as tattoos, scars, or religious or tribal physical mutilations or blemishes (circumcision or branding, for example) would require consideration and perhaps negotiation.

How the Surgery Might Work

A virtual army of medical professionals would be recruited to complete a BHT. These professionals must be willing collaborators, which could be difficult given the ethical questions surrounding the procedure, and fully understand any liabilities they may incur. Alex Watt (2015) projects that about 100 surgeons, anesthesiologists, and other specialists are expected to work in a closely choreographed and rehearsed schedule with numerous nurses, medical technicians, and other medical and administrative staff members. If the surgery is successful, the surviving patient would need extensive rehabilitation services. To aid recovery, a patient would require access to physical, occupational, and psychological therapists, and social workers would be necessary to manage post-operative supportive services such as securing housing, transportation, money management, and emotional support systems.

To start, the primary directors of the experiment would divide surgical personnel into two teams. As one team works to remove the head of the body donor and prepares the body to accept the donated head, the second team removes the head donor's head and prepares it for reattachment. The window of time to complete the first stages of BHT is both small and critical, and the significance of following the sequence of steps with exact precision cannot be underestimated. To avoid ischemic damage, the body temperature of the two patients would have to be reduced to 10 degrees centigrade (50° F). The patients' necks would then be prepared for separation from their bodies. Arteries, veins, and muscles would be identified and labeled for future reconnections, and the trachea and esophagus prepped for transfer via cervical incisions. Then, according to Canavero, the vertebral columns would be transected at the C3 and C4 level, and both spinal cords bisected using an ultrasharp, diamond-edge surgical blade specially engineered for minimal force application to reduce fraying.

At that point, Canavero contends that the "live" head would be separated and exsanguinated to minimize the risks of coagulation. Then using a specially designed crane, the head would be transported onto the now head-less donor body. Timing is essential at this stage of the operation. The brain, which uses about a fifth of the body's oxygen supply, can only tolerate about three to four minutes without oxygen before permanent damage sets in. Once the head is detached from its body, surgeons must work with streamlined efficiency to re-oxygenate the brain before impairment occurs. It is likely that surgeons would connect the head to the donor body's circulation system with tubes that link its carotid arteries to the donor head before it is fully attached to reduce risk of ischemic injury. Most likely, a heart-lung machine would be used to minimize the risk of injury during the transfer.

Once the brain is oxygenated, the spinal cord ends would be fused together using a plan code-named GEMINI (Canavero 2015b). This, too, has a crucial time constraint. The ends of transected spinal axons remain intact for only about 10–20 minutes before they begin to degenerate (Canavero et al. 2016a). Canavero's proposal suggests coating the ends in polyethylene glycol (PEG), a glue-like substance believed to stimulate neuron growth. PEG would be infused intravenously into the patient's bloodstream to fill in neuronal gaps to promote axonal healing. In addition, Canavero suggests implanting electrodes on each side of the "point of fusion" to strengthen and stimulate nerve connections and inserting a negative pressure device to create a vacuum around the connections to promote spinal cord fusion (Canavero et al. 2016a). At the heart of GEMINI is the notion that the PEG coating and electrodes will spur growth in the spinal cord's gray matter "sensory-motor highway" (Canavero 2015b). He further states that "gray matter neuropil will be restored by spontaneous regrowth of the severed axons/dendrites over very short distances at the point of contact between the apposed cords" (Canavero 2015b). Neuropils are the dense network of neural structures such as axons and dendrites, glials, and fibrils in the gray matter of the central nervous system.

As the end of the surgery approaches, the spines of the head and body would be fused with a titanium plate installed with vertebral screws at C3-5, which is consistent with common surgical practice, and the anastomoses of vessels.

After the head is stabilized with an elaborate screw and rod system, the final steps would include connecting the remaining tissues, including muscles, and closing the skin from the body and the head. According to the plan, the now transplanted head and body would continue in sedation under an induced coma that has been variously estimated to last from three days to four weeks. Maintaining the coma, according to HEAVEN, would prevent movement and allow time for the spinal cord to heal. Lastly, an immunosuppression regimen, which would likely begin before the surgery, would be continued as the patient is monitored for signs of rejection.

The surgery would likely last about 36 hours (Watt 2015).

Assessing and Identifying the Benefits and Risks of BHT

Just by reading this abridged review of what BHT entails,[3] numerous risk junctions emerge as danger zones in which a chance of failure or unexpected turn of events is particularly high. In addition, assuming that the surgery works and the patient survives, the post-operative state of the body recipient's mind is unpredictable. We have no way of knowing how a person will adjust psychologically, and we do not know how the brain and the new body will integrate into a normal functioning whole.

Human consciousness and personality are simultaneously resilient and fragile. People can indeed survive unbearable distress, but they can easily experience severe identity crises and harmful emotional states following a trauma. It seems rather clear that BHT constitutes a physical and emotional trauma. The severed spinal cord, the almost certain post-operative pain, and the healing of muscles occurring concurrently with the shock of the brain trying to integrate with a completely new body could create a disturbance unlike anything ever experienced by a human being. The severe physical distress will certainly accompany emotional trauma.

How this trauma plays out cannot be explored experimentally prior to the first BHT attempt. How the brain integrates with a new body is unexplorable with animal, cadaver, and brain-dead studies. We must infer what could happen to the mind and deduce the likely challenges a patient should expect.

The remainder of this chapter reviews the benefits and risks inherent in the BHT process.

---

[3] Many steps of the BHT proposal have been excluded from this review. For example, patients would require ventilation and be tracheotomized and intubated to maintain stability. Body temperature and oxygen levels would be watched closely, and electrocardiogram and electroencephalograms would monitor essential life functions. These steps, among others, are necessary to achieve surgical success but are not necessarily unique to BHT, measures of high risk, or likely to impact the condition of the patient after the surgery.

Medical Benefits

The intended consequences of BHT are the remedy of terminal disease and the continuation of the life of the patient. BHT is not a cure *per se*, but a strategy to neutralize illness by removing the diseased and dying body and replacing it with a healthy one. The medical benefit therefore is straightforward: BHT is proposed as a life-saving composite tissue allotransplantation.

In a recorded interview, as mentioned in Chapter 1, the architect of BHT has added two additional potential benefits to the experiment (Canavero 2022a). One is immortality. The idea is that an individual, as defined by the memories, personality, and identity stored in the brain, could continue to live by getting a new body as the first (or previous) body becomes disease ridden, infirm, or old. Transplanting the head onto a cloned body was another option Canavero posed.

This idea is reminiscent of Michael G. Coney's 1973 novel, *Friends Come in Boxes*, in which individuals' brains were transplanted into new bodies when they turned 40 years of age. While awaiting a new body, brains were stored in specialized boxes that fed them nutritional supplements and oxygen and allowed them to hear and speak but not see or move. When a new body became available (six-month-old children provided them), the brain was surgically placed into the baby's head, a procedure that could be conducted repeatedly every 40 years, providing the embodied person obeyed the law. The intent of these transfers was to reduce the population, which had grown out of control and was threatening the sustainability of global resources. The book investigated how public policies, in this case population control programs, disrupt the lives of individuals and families. Coney also raised questions about the misuse of technology by an autocratic state. The author was not medically trained, so he was not particularly interested in how brain transfers would really work. Perhaps more importantly, he also avoided what happens to a human brain as it ages.

Should physicians use BHT to create immortality, they would also have to figure out how to prevent changes in the brain that come with time. The brain would not refresh itself with each new body in which it is housed. Just like every other organ or body tissue, the brain ages. With time, grey matter shrinks, which can impact learning and other complex mental activities, communication between neurons is less effective in certain regions of the brain, and blood flow often decreases while inflammation increases (National Institute on Aging 2020). While older individuals retain the ability to learn new tasks, losing cognitive speed and some memory functions especially with words is normal to an aging brain. And there is more. As the brain gets older the incidence risk for stroke, white matter lesions, and dementia rises, and the levels of neurotransmitters and hormones change (Peters 2006). There is no medical or psychological reason to believe a new, younger body will automatically rejuvenate an old brain and reverse the aging process.

The cloning option is also rife with medical and ethical problems. For one, cloning does not solve the aging brain dilemma. The only way to make that work is the age-old science fiction device of transferring the contents of one brain into a synthetic or robotic brain. The ethical limits on immortality through cloning are equally imposing. If Person A's body is cloned to create Person A-Clone, would not Person A-Clone have a brain? And would not that brain be a dynamic organ capable of learning, retaining memories, and developing a sense of self? By definition transplanting a head onto a cloned body means murdering the personhood of the genetic duplicate who exists as a full-fledged human being with psychological functioning, the capacity for learning and self-reflection, and legal rights to liberty and the pursuit of happiness. This proposal is akin to Coney's science fiction plot device in which the brain transfers went into the bodies of six month-old children, individuals whose legal right to life was judicially and normatively abandoned because of the immediate societal need to control population size.

Immortality is neither a worthwhile motivation to pursue BHT nor a serious medical research question at this time. Philosophers have long discussed immortality as a thought experiment, and many have concluded that existing forever may contradict the essentialities of living a full and meaningful life (Parker 2022). Bernard Williams (2016), for example, contends that immortality would undermine our drive to create fulfillment in life. Death gives meaning to our existence because knowledge of our mortality inspires and motivates us to accomplish our desired goals and fulfill our aspirations. Immortality would make goals rather pointless and would induce boredom. As Samuel Scheffler (2013) posits, mortality reminds us to cherish life. He further argues that living forever would weaken the value of achievement, love, creativity, and humor simply because that which gives us joy in life would no longer be necessary. We cherish love in life because in death we know we will lose it.

Such arguments are compelling, but it is the impracticality of achieving immortality in the way Canavero envisions via head transplantation and cloning keep us at a safe distance from testing the hypotheses of eternal life on Earth.

As with every new technology, there are unintended consequences of immortality. First and foremost, would be over-population. Should people live forever and continue to reproduce, the population would be too large to sustain. People would have to be denied the right to have children. Children are conceived for different reasons. For many, especially the poor, children are needed to contribute to the household's income. For others, children are equated with love, and as the saying goes, "love triumphs all". How will society incentivize or coerce people not to have children? If Williams and Scheffler are right, perhaps immortal people will find sex boring; but the body may have something to say about that. Sexuality is a potent and driving bio-psycho-social force in people during their most fecund years. Second, how is immortality technology dispersed and to whom? What will those rules be, and who will create them? Would the technology be available to everyone, regardless of their social prestige, status, or ability, or to just a select few who are judged to be of greater social value? It is likely not every person would get to live forever, so some are allowed immortality while others must die and perhaps without having children, all for socially constructed reasons. It would quickly evolve (or devolve) into a new form of eugenics. Political and social upheaval would likely follow and be so riotous that implementing immortality would defeat the purpose of living cooperatively, at least as we do now. In Coney's fictional world, immortality is maintained through totalitarianism, a vision that could prove a presage if BHT, cloning, and eugenics were to merge.

Perhaps future technology derived from expanding our understanding of the role of genetics in aging will make immortality closer to a reality, regardless of what it may do to the human spirit. But for now, short of any miraculous medical breakthroughs, entertaining thoughts of immortality are impractical, and we must resign ourselves to the confines of our temporal impermanence and leave the possibilities of eternal life in the hands of theologians and fantasy writers.

Another possible use of BHT, as Canavero presented, is that it would allow humans to live on Mars. The radiation exposure explorers and colonists would face in route to and once settled on the Red Planet would cause a number of illnesses that would likely shorten life expectancy. Chronic illness, therefore, would interfere with social stability in the new colonies and require us to restock the Martian population more frequently than would otherwise be necessary. BHT would solve that problem, according to Canavero. It is not clear how on Mars those new brain-dead bodies would be secured or how the legalities and practicalities of creating and then ending the lives of the cloned colonists would be managed.

While such a story might make for entertaining reading in a novel, this notion approaches the absurd when presented as legitimate science, at least within the confines of current social and ethical sensibilities. The idea of replicating a person for the purpose of perpetuating the brain of that person implies killing the clone and therefore breaching the most sacred of human and medical ethics: the promotion of individual integrity and human life itself.

The Risks of Body-Head Transplantation

Let us return to the more likely scenario of BHTs in which the surgery would be employed to save the life of a dying person. Fantastical uses as described in the preceding paragraphs obscure the real intentions of BHT proponents and divert our attention from how complex this experiment actually is and the many possibilities that something could go terribly wrong.

A number of unintended outcomes at risk of occurring in the BHT blueprint present themselves. It would be simpler if we could divide the risks into neat categories such as medical and psychosocial, but such a typology would yield an inaccurate representation of the possible outcomes of BHT because they are not discrete phenomena. The risks are complex and overlapping and are neither isolated nor discrete. They are like falling dominoes—one problem begats others.

This is especially true when considering the psychological aspects of BHT. Medical and psychosocial phenomena intersect, each acting in concert with the other. When a disruption occurs in one, it impacts the functioning of the other. BHT perhaps constitutes the greatest interruption between normal mind and body union that we can imagine. Dissecting the spinal cord, preventing rejection, and the neurological implications of a brain and body attempting to create a new homeostasis but are unfamiliar with each other will have potentially serious consequences on social and behavioral aspects of personhood for the surviving head donor. Introducing a brain to a new body constitutes both medical and social-psychological problems. In a sense, they are two sides of the same coin: where one goes, the other follows.

Death

To put it directly, not surviving the BHT surgery is a real possibility. Animal studies have demonstrated a very low survival rate, measuring survival in hours and days. The avenues for death are numerous, and at this time data would lead us to believe that not surviving is perhaps the most likely outcome of an attempted BHT. The conduits to death are tissue rejection, failure in the decapitation and reattachment process, surgical complications, and the absence of an exit strategy.

Preventing the body and head from rejecting each other is one of the key components of a successful BHT transplant. Problems linger for immunosuppression. First, the published animal studies have paid little attention to preventing rejection, so there are no data on the probability of this risk. No tissue of the mass of a human head (or body) has ever been transplanted which means medications and dosage are uncertain. In addition, the brain has never been transplanted, and while the brain is protected by the blood-brain-barrier (BBB), it is not known what will happen to the brain in terms of rejection should a leak in the BBB occur. Transplanting a brain wanders into unknown territory.

Second, to minimize surgical impediments in the decapitation and reattachment stages, Canavero and Ren (Ren et al. 2017) conducted a cadaver experiment in China in which they decapitated two bodies and reattached one head onto the other body. They pronounced the experiment a success, but the study amounted to little more than a rehearsal of the mechanical steps since the two patients were not alive. Success in terms of survival is undeterminable from this study. A cadaver study elicits no information on rejection, pain, body and head integration, infection vulnerability, blood and oxygen supply, neurological regeneration and functioning, or psychological effects. Such a study does not predict if a patient will live.

Third, the list of possible complications of a surgery as complex as BHT is quite long (Ilitis 2022). A miscalculation at every step of the way can be catastrophic. Oxygen depletion to the brain, excessive blood loss, a failure in neurological functions, infections, and restoring respiration are just some of the surgical risks physicians must prevent during the course and immediate aftermath of the surgery. Another surgical risk is the medically induced coma designed to allow the spinal cord to fuse and heal. Canavero (2015b) proposed that the coma following BHT could last as long as a month. Induced comas carry risks for a wide range of side effects, but the effects of protracted comas are not entirely predictable. It is generally believed, that the longer a coma lasts, the higher the likelihood of brain injury and blood clots, among other possiblities.

Decapitation obviously implies killing both body and head donor. In BHT, it is presumed that this death is temporary for the body-receiving head. Restoring oxygen to the brain is vital to avoid brain hypoxia and reconnecting completely severed spines and spinal cords has never been done on living patients. Thus far experiments on animals demonstrate that these two procedures produce limited neurological functioning, but again, the long-term implications for survival, unassisted respiration, ambulation, fine motor control, and psychological adjustment are unknown.

Other composite tissue allotransplantations have exit strategies in the event of catastrophic failure. In hands and feet transplants, if a significant surgical problem that impairs functionality or if tissue rejection cannot be controlled, the new tissues can be removed, assuming the problem is addressed before the problems become systemic and the whole body is affected. Removed transplanted hands and feet can be replaced with a prosthesis or another transplant. In face transplantation, where removing transplanted tissue is more difficult, the worst case scenario is that the patient would die. But short of death or irreversible illnesses caused by immunosuppression drugs, physicians could conduct a second transplant, which has happened on at least two patients, or replace the transplanted tissue with skin grafts and plastic surgery reconstructions. This outcome is of course preferable to death but carries its own problems in that the patient would remain severely disfigured and would have experienced the physical and emotional trauma of face transplant surgery for nothing.

No comparable exit strategy exists for a patient undergoing BHT. If technical complications or uncontrolled tissue rejection cause the transplant to fail, the head and brain could not be sustained on life support for sufficiently long periods to identify another donor body. There is no reason to believe a second trial would result in a better outcome. Such a failure would imply certain death for the patient (Furr et al. 2017). Once decapitation is conducted, there is no going back.

Spinal Cord Fusion and Paralysis

The blueprint for BHT requires severing the spinal cords of both the body and head donors. Because such a dissection has never been undertaken for the purposes of reconnecting the severed ends, this phase of the operation is a target of much criticism. Generally, physicians believe that spinal cord fusion is not possible. For these critics, the best a BHT patient can hope for is tetraplegia, complete paralysis. If modern medical technology has been largely unsuccessful in healing spinal cord injuries, physicians question how reattaching completely severed spinal cord ends of two different people would work.

That said, a number of studies have provided a smidgeon of encouragement that Canavero's GEMINI protocol could work (Ren et al. 2019). In hundreds of mouse, rat, and dog BHTs and experiments reconnecting severed spinal cords, Ren and his colleagues have improved upon surgical techniques that reduce the likelihood of cerebral ischemia and enhancing breathing and circulation (Ren et al. 2014; Ren et al. 2015; S. Ren et al. 2019). One study (S. Ren et al. 2019) funded by the National Natural Science Foundation of China dissected the spinal cords of 12 dogs and compared the neurological recovery of those whose cords were reconnected using PEG and those dogs in which PEG was not applied. The PEG-treated canines showed signs of recovery and could move their limbs within three days. The control group, those without PEG, demonstrated no neurological healing. Research groups led by South Korean veterinarian CY Kim (2016) and Harbin University neurologist Shuai Ren (2019), both research collaborators of Canavero and X. P. Ren, reconnected a dog's spinal transection using the GEMINI process of fusing spinal cord endings with PEGs. In the first study, according to the authors, the dog regained about 90 percent of its normal sensorimotor functioning three weeks after the experiment,. In the Shuai Ren-lead experiment, two dogs whose spinal cords were transected in a manner consistent with the GEMINI protocol were followed for two years. Their recovery was evidenced by measured axonal sprouting "across the fusion interface" (S. Ren et al. 2019). Similar research by Wang in 2020 on rats also found that PEG scaffolding improved locomotor functioning eight weeks after their spinal cords were transected.

The concerns for spinal cord fusion are not limited to attaining motor skills and ambulation. Creating the neurological pathways to maintain automatic breathing worries many physicians who believe that not only is tetraplegia a likely outcome, but living on a respirator is also probable (Van Assche and Pascalev 2018a). Ren's experiments seem to lower the risk level for respiratory failure, but long-term survival has yet to be demonstrated and the probability of recovery, as indicated by the number of animals who survive with any neurological functioning, remains too low to justify extrapolating to human subjects. No animals subjected to decapitation and transplant appear to have been restored to full neurological normalcy.

One review of 15 research articles revealed many advancements in spinal cord repair using GEMINI-related protocols that may have implications for BHT (Gotebrowski and Gotebrowski 2018). Although the experimental results fell short of confirmation of concept, they do suggest that further research is warranted to determine if deep profound hypothermia and PEG fusogens can solve the problem of spinal cord repair and fusion.

Not all are convinced. Polish scientists Piotr Zielinski and Pawał Sokal (2016) propose a theoretical argument against complete spinal cord anastomosis. Their idea assumes that axon growth is strongly inhibited in mammals. They state that peripheral nerve axon regeneration seems to form random patterns of their target reconnections and hypothesize, using their words, that due to the laws of entropy or irreversible information loss, full spinal cord restoration is not possible. If this is indeed what would happen in a complete spinal cord transection, each axon in the spinal cord has to grow and reconnect with exactly matching targets below the dissection. The odds of this happening, according to Zielinski and Sokal, are slim. If reconnections are indeed random for the approximately 2,000,000 pyramidal axons, then there are 2,000,000! permutations, making the chance a pyramidal axon connecting to its "adequate target" 1/2,000,000!.

It is generally believed that aligning the spinal tracts is next to impossible. If they do not align, then both ascending and descending signals would be cut off and would go no further than the transection. If neural endings were to overlap, a signal may continue though its strength will be weakened, leaving patients with some sensory function and a diagnosis of untreatable incomplete spinal cord injury.

Canavero (2013) asserts, in contradiction to Zielinski and Sokal, that axonal alignment is not necessary for union. Instead, he proposes that spinal cord gray matter inter-neuronal networks would remain intact and functional if stimulated with low-level electrical stimulation. The contradictory hypotheses posed by Canavero and Zielinski and Sokal have yet to be sorted out in animal research to warrant confidence to apply the HEAVEN approach to human anatomy.

Though Ren and colleagues (2015) have maintained blood flow to experimental mice for as long as six months, it is worth repeating that survival for the laboratory animals in Ren and Canavero's experiments is generally measured in hours and days and that complete neurological functioning after a transplant has not been reported. Outside of Canavero's and Ren's own networks of researchers, there are few who have faith in the GEMINI concept, and researchers generally reject extant documentation that shows that functional spinal cord fusion and post-operative pain control are feasible (Lilja-Cryon et al. 2018).

Doubt lingers that animal studies using polymers are not generalizable to human spinal cords (Barker et al. 2018, Gerasimenko et al. 2015). Indeed, the evidence for transferring the results of these animal studies to humans is scarce (Lamba et al. 2016). Dogs, cats, rats, mice, and monkeys have different spinal cord circuitry and regenerative capacities than humans; therefore, it is not certain if any animal success is transferable to humans. Even scientists who are generally sympathetic to the notion of BHT and Ren and Canavero's proposal are skeptical that spinal cords can be fused. Still, supporters such as Gkasdaris and Birbilis (2019) believe in the theoretical premise of GEMINI and that BHT is "worth attempting".

As Furr, Hardy, Barret, and Barker (2017) have reported, other streams of research on human spinal cord injuries offer promising technologies to restore neurological functioning after a spinal cord injury. These approaches can be generally placed in one of three categories: 1) delivery of neural-derived cells into the site of injury or bypassing the injury using peripheral nerve grafts, 2) modification of the CNS environment, and 3) electrical stimulation. An example of the first type consists of using stem cells harvested from the olfactory bulb (olfactory ensheathing cells), which when implanted into the site of non-scarred injury have been shown to act as a bridge for the distal and proximal ends of the injured spinal cord through the transected tissue enabling axons to regenerate and reconnect (Keyvan-Fouladi et al. 2003). A recent clinical case-report using this approach described partial restoration of ambulation in a patient who was confined to a wheelchair following complete cord transection (Tabakow et al. 2014).

Another approach used in several different animal models and a few clinical cases consists of positioning peripheral nerves to bypass the spinal cord injury. This method has been shown to restore motor and sensory function albeit incompletely (Oppenheim et al. 2009).

Finally, electrical stimulation may promote axonal regrowth following spinal cord injury in both animal models (Borgens et al. 1990; Borgens et al. 1993) and in clinical settings (Brissot et al. 2000). In tests with several patients with complete lower extremity motor paralysis, electrical current and training were applied to their spinal cords. The treatments re-established functional connectivity among the neural networks that link the brain and the spinal cord. Afterwards patients attained voluntary movement in their lower extremities (Gerasimenko et al. 2015).

In another approach using electrical stimulation, scientists have implanted a device in the motor cortex of the brain's frontal lobe that records brain activity associated with locomotion. The primary function of the frontal lobe is to generate signals that direct the body's movements. Brain activity detected by the device is transmitted to a computer that interprets, reconfigures, and wirelessly transmits the signals to a second device that stimulates nerves in the lower spinal cord, away from the injury, that in turn stimulates muscles in the limbs responsible for locomotion. Using this technology, researchers have restored almost normal walking in previously paralyzed primates (Capogrosso et al. 2016). It must be noted that these treatments typically attend to spinal cord injuries, not fully transected spinal cords.

Because spinal cord fusion and complete neurological recovery have yet to be achieved in animal experiments and that experimental transections and repair on living humans are impossible studies to conduct, the general medical community is doubtful BHT will restore full neurological functioning as intended. For skeptics, the best scenario at this point is that a patient might survive the surgery but be left in a state of tetraplegia. The best of today's technology predicts that tetraplegia would continue for an uncertain duration, if not permanently. Patients who survive would likely express relief that their life-threatening illness has ended; however, in time they may experience the psychosocial consequences, such as low quality of life and depression, that often accompanies tetraplegia (Rousseau et al. 2013).

The idea that two disparate spinal cords can be joined and once healed will allow normal functioning remains a high risk "flag" for patients considering undergoing BHT. Unlike injured peripheral nerves that can regenerate, it is generally believed that the central nervous system, once wounded, does not grow back and start functioning again.

Memory Loss

Once the brain hits the three to four minute mark without oxygen, its cells begin to die. The hippocampus, which is the area largely responsible for memory functions, is highly susceptible to anoxic injury, which can severely impair patients' ability to understand what they have experienced and even know who they are. After surgery, they could easily ask, "What am I doing here?"

With an anoxic injury, patients' inability to reconcile their now unknown past with the present means that the subjective experience of transitioning from their desperate health condition and psychological state before surgery to a new and unfamiliar present is ambiguous to both patients and surgeons before surgery (Suskin and Giordano 2018). How do patients and surgeons decide if the surgery was a success if patients have no memory of life before the transplant? What would the quality of life be for someone with severely impaired memory and other brain deficits? And what resources are necessary to keep brain injured patients stable and healthy with an acceptable quality of life? What does quality of life look like for patients whose memory loss prohibits them from comparing their life experiences before and after the surgery?

Minimizing Tissue Rejection

Successful transplants rely on suppressing the host body's immune system so that the body does not see the introduced organ or tissue as foreign. The immune systems' job is to attack foreign "invaders", which in a healthy person it does successfully and efficiently. When a person gets a transplant, a friendly "foreign agent" is admitted to the system to improve system functioning and health. But the immune system's "intel" doesn't understand that agent as friendly. Unfortunately, the immune system has "tunnel vision" and cannot distinguish vicious enemies, such as viruses, from desired "friendly" alien collaborators such as transplanted organs.

Thus, for a transplantation surgery to work the immune system has to be suppressed so that it does not attack or reject the transplanted organ or tissue. Suppressing the immune system comes with high danger potential because it weakens the body's defenses and exposes transplant recipients to health problems they may not otherwise have faced, and many of the complications of immunosuppression are serious and life-threatening. Patients on immune-suppressing medications have increased risk of such disorders as: Epstein-Barr virus-related B-cell lymphomas; various cancers including squamous cell, lung, colon, and non-Hodgkin lymphoma; opportunistic infections such as herpes simplex virus, cytomegalovirus, pheumocystis jirovecii (a fungal infection that can cause death if untreated), and various bacterial infections (Iske et al. 2019).

Despite these risks, the sophistication of today's immunosuppression regimens makes transplantation highly accessible, safe, and reasonably durable. After a transplant surgery, immunosuppression medicines are carefully monitored and are frequently tweaked to find the right dosage to match patients' needs, repel rejection episodes, and minimize future rejection events.

It is expected that the majority of cells that would reject BHT will come from the donor body and attack the face (skin), the spinal cord, and other structures, such as the pituitary gland, that reside outside the brain's own protective network, the blood brain barrier, or BBB.

BHT poses unique problems for immunology and managing rejection. The success of such an unusual allograft, where the donor and the recipient can reject each other, depends on prevention of complex immunologic reactions, especially rejection of the head by the body (graft-vs-host) or probably less likely, the possibility of the head rejecting the total body allograft (host-vs-graft). The immunologic difficulties facing BHT are enormous, especially since rapid nerve and cord connections and regeneration are not likely, leaving the body's immune response confused (Hardy et al. 2017).

Managing rejection can be trying when skin is transplanted. As we saw earlier with composite tissue allotransplantations, compared to transplanted solid organs, controlling rejection is trickier when skin is involved, though rejection events can be detected sooner in transplanted skin since it is easily visible, as compared to organ rejection which is often not detected until the patient has become ill.

Transplanting skin is not the only cause of concern for BHT. Another problem of immunosuppression in a BHT is the impact on the transplanted brain, which is said to be "immunologically privileged" because it is encased in its shielding BBB. The BBB is a protective network of tightly arranged cells located between the brain's blood vessels and the cells and other components of the brain. Cells that mediate immune reactions will not cross the BBB. The BBB defends against disease-causing pathogens and toxins that may be present in blood, and as long as the BBB remains uninjured, the donor body should not recognize the donor brain as foreign tissue.

The brain's "privileged" position, however, is not without compromise (Hardy et al. 2017; Szalavitz 2002). The BBB is not impenetrable. An ischemic injury may cause a breach in the protective "wall" and calcineurin inhibitors, which are sometimes used in immunosuppression, are known to have neurotoxic effects. The BBB can also be missing in some cases (Barker et al. 2018). Should any of these scenarios unfold, the body's immune system could reject the brain, which likely would be catastrophic.

Perhaps the best strategy for successfully suppressing the immune system is a pre-transplant immunologic manipulation of the body donor and the recipient. Several standard immunosuppressing drugs such as cyclosporin A or Tacrolimus, and calcineurin inhibitors would be used. Sirolimus, another commonly utilized rejection prevention medication, is also known to aid in recovery of nerve and spinal cord function and BBB leakage (Liu et al. 2014; Kawai et al. 2008).

Relying on drugs alone may prove insufficient to assure patients of reasonably few and manageable rejection events, and alternative forms of tolerance may be required (Barker et al. 2018). For example, establishing chimerism, where genes from both donor and recipient exist in the unified body, has shown promise in eliminating transplant patients' dependence on drugs and the risk of their side effects (Spitzer and Sachs 2022). Chimerism, often called the Holy Grail of immunology and transplant medicine, already has been achieved in a handful of patients in very specific conditions (Kawai et al. 2008).

For BHT, one strategy for creating a chimeric state is populating the recipient with donor bone marrow cells. The donor body would have to be irradiated, shielding vital organs, to deplete it of donor T cells, B cells, plasma cells, and a variety of cytokine-producing activated macrophages (Barker et al. 2018). After irradiation, which is also proposed by Ren and Canavero (2017a), the body becomes an "empty vessel" immunologically but can be reconstituted with bone marrow from the (head) recipient's body, thus establishing a chimeric state. The removed body of the head recipient would provide bone marrow cells for "booster" injections, as well as cells from various organs for any needed organ repair.

Chimerism is proving possible and showing promise in the early stages of its development. Nevertheless, it is far from established practice and its reliability does not sufficiently provide a sound base for a BHT. In addition, there are scientific and ethical constraints. Maintaining the deceased donor body for up to four weeks may prove emotionally difficult and contrary to the wishes of the donor's family who may agree to the donation but be hesitant to allow the time needed for all the manipulations necessary to achieve chimerism for the recipient head (Barker et al. 2018).

Physicians generally agree that immunosuppression of the body and head would begin before the surgery and that it would require irradiating the body's bone marrow and possibly removing the spleen. These complex procedures would occur in tandem with standard drug therapies for minimizing tissue rejection. The risks in conducting these treatments are substantial.

Controlling immune responses is crucial to the success of BHT, as it is with any transplant. In the ideal scenario maximal elimination of donor cellular components by targeted sub-lethal irradiation to deplete antibodies, "standard" immunosuppressive therapy, and even splenectomy and its replacement with the recipient's bone marrow cells could reduce or even eliminate rejection episodes. Initial treatments of rejection episodes can use established successful vascularized composite allograft rescue protocols such as tweaking medications[4]

---

[4] Vasculature refers to a net-work of blood vessels connecting the heart with all other organs and tissues of the body.

(Barker et al. 2018). Not preventing rejection of the new head or the body will undoubtably be fatal (Hardy et al. 2017).

Pain

A BHT surgery is expected to leave the surviving patient in racking pain, making a plan for pain control of great necessity. With the exception of Jerry Silver's report on one of White's monkeys, as presented in Chapter 1, animal studies cannot totally inform us on the amount of pain the surgery causes, and those who have conducted the experiments more recently have not fully addressed how much pain the survivors will suffer and what to do about it.

Pain is an enigmatic and subjective experience which means that it cannot be reliably measured by an instrument or direct observation (Wideman et al. 2019). The best we can do is rely upon patients' own narratives, or descriptions, of the pain they experience. As Margo McCaffery (1968) famously said many years ago, pain is what people say it is and exists whenever they say it does. To add to the subjectivity of pain, we know that individuals interpret and express pain differently. Individual characteristics and social background, especially social class, racial, and ethnic group membership, influence how individuals interpret and report the pain they experience. Many factors such as socio-cultural norms and personality traits (traditional British stoicism, for example) help account for an individual's narrative expression of pain.

Relying upon patients' own reporting of their pain assumes that they are able to express it. If patients have difficulty speaking or moving, they cannot report their level of pain and where the pain originates in their bodies. If they are confused or their neurological connections are not working properly, patients will have difficulty identifying and reporting their pain.

Canavero and Bonicalzi (2016) say that central pain is "possible", but everyone who has commented on BHT and mentioned pain agree that the spinal cord transection and reattachment would undoubtably cause terrible misery for the BHT patient.

More specifically, pain would originate from cutting the spinothalmic tract, which is a sensory neurol pathway that carries nociceptive pain, temperature, touch, and pressure from the skin to somatosensory area of the thalmus. The thalmus is like an information relay station that receives all information from the body's senses (except smell) and then forwards them to the cerebral cortex for interpretation (Gkasdaris and Birbilis 2019). Severing the spinal cord will lead to a severe form of central pain syndrome (CPS) that Van Assche and Pascalev (2018a) believe would be excruciating.

CPS can be specific to a particular part of the body such as feet or diffused throughout the body, and it can range in severity from moderate to severe. Being touched, moving, temperature changes, and even emotional experiences can worsen the intensity. Numbness and/ or burning sensations are common, especially in the hands and feet. The common characteristic of CPS is pain in the form of feeling "pins and needles", pressing or aching pains, or acute bursts of nerve pain. CPS is almost always constant.

The experience of central pain syndrome varies by what causes it: CPS can stem from strokes, multiple sclerosis, tumors, and Parkinson's Disease. But trauma to the spinal cord or central nervous system may cause the most severe central pain. Radi Masri and Asaf Keller describe patients' pain with insults to the spinal cord as suffering from "excruciating, unrelenting, chronic pain that is largely resistant to treatment" (2012:74).

BHT, therefore, is almost certain to cause tremendous and unbearable pain to surviving patients. Canavero and Bonicalzi are cognizant of the potential for central pain:

> One dreaded consequence [of BHT] is the possible onset of cord central pain (or central neuropathic pain) …. In such eventuality, the patient would receive a new body, but go on suffering for the rest of his/her life a most excruciating chronic pain (2016: 271).

Nonetheless, Canavero has proposed a method to stop the agony that BHT promises:

> Known to medicine since 1891, CP [central pain] has remained a mystery for over a century until Canavero proposed the Dynamic Reverberation theory in 1992, which led to a rational treatment. Should the first head anastomoses develop CCP [central pain], a cure, however *experimental* — is available (Canavero and Bonicalzi 2016: 271).
> [emphasis added].

Canavero and Vincenzo Bonicalzi, who have published together for many years, first presented their theory on central pain in the early 1990s and have continued to promote it in another article and a book published by Springer in 2018. While I encourage readers to refer to these publications for the details, the technique essentially focuses on the use of ultrasound to neutralize and "cure" central pain. Ultrasound produces heat, and BHT physicians propose to use it to burn and kill the neurons that transmit ascending pain sensations. While this proposal could indeed terminate a patient's pain, the patient may never experience pain or other sensations below the ablation again. What is important for our story is that the treatment remains theoretical, yet in Canovero and Bonicalzi's own words, holds "clear promise" (2016: 274).

Many physicians and bioethicists have warned that a BHT survivor's quality of life could be worse than their existence living with a terminal disease before the surgery. The uncertainty of whether central pain could be managed is certainly one risk that patients would have to consider before providing informed consent. At this time, it is reasonable to conclude that the risk of severe chronic pain remains high.

The Loss of Identity and Psychological Health

The subject of the impact of BHT on the psychological well-being of surviving patients often arises among skeptics. Only the surgery's most devoted advocates aver that uniting a new body with a brain-encased head will have little or no impact on post-operative emotional health. Many opponents of BHT believe that as the new body strives to connect with its new brain, patients' sense of self will blur if not distort and lead to their feeling confused as to who they are. Others forecast more dire consequences: that the brain and body disconnect will render a patient completely mad.

A long list of questions posed by critics linger largely unattended by BHT proponents. Skeptics wonder: Who will awaken from the anesthesia? Will that person be the person contained within the brain, or will there be a change in personality and self-identity? Will the survivor be a "new person", a third entity that is neither donor nor recipient, possessing a sense of self that did not exist before the surgery? Will the material body intersect with the non-material mind in a way that changes how people see and understand themselves both psychologically and within their social environment? Do mental illness and health totally occur within the brain or does the body contribute to psychological well-being?

Because these questions hold legitimate scientific merit but have not been seriously addressed by BHT advocates, we must consider post-operative psychological health as a genuine risk area for patients. Assessing psychological well-being before the surgery, while important and necessary, is not the same as predicting the impact that BHT may have on psychological health. A pre-operative evaluation is a cross-sectional assessment of patients' psychological status at the time the tests are conducted. They may determine patients' emotional stability and cognitive ability at the time of the assessment, but testing's predictive powers are subject to error. Most importantly, there is no method by which mental health professionals or neurologists can measure how a brain will adjust to a new body, which could have significant implications for mental health.

Since there are no valid or reliable methods that could predict how a person will adjust to having a new body, comparisons have been made between face and head transplantations. Proponents have suggested that a head transplant is essentially an oversized face transplant. If they are essentially equivalent, the two procedures should have similar psychosocial sequelae—face transplantees and body recipients would likely have the same experiences after their surgeries. This hypothesis is indeed intriguing, and there are some similarities in the histories of the two procedures. Before the first faces were transplanted, bioethicists were concerned that changes in appearance might inflict harm on transplantees' identity and sense of self, just as they are with BHT. In the late 1990s and early 2000s many alarms were raised that recipients would be unable to handle having someone else's facial appearance and that foreign skin in such an intimate location would be hard to accept. A new outward self would translate to a confused inward self, making patients uncertain as to who they are if they do not look like themselves any longer.

Research, however, found that this was not necessarily how face transplantees reacted to their new appearance. The mental health of some, but not all, face patients has been analyzed and reported in scholarly journals. From these accounts we have learned that face recipients' self-esteem, social integration, and overall mental health have indeed improved after their transplants (Chang and Pomahac 2013; Nizzi et al. 2017). We should not be surprised to see that a new and more normal appearance improves overall emotional and relational health given the importance of our faces for self-esteem and the harsh impact that disfigurement has on body image and social relationships. As they adjusted to their new physical self, face transplant patients' probably compared how they looked after surgery to their severely disfigured appearance that was the result of a trauma or disease. Knowing they have a more normal looking face probably brought relief to the emotional suffering they likely felt when their appearance was dramatically different from most other people's.

The experience of face transplantees is not so different from solid organ recipients. One of the successes of transplant medicine is that patients almost universally accept their new organs and tissues and integrate them into their own embodied self. Thinking of the donated organ as a "gift of life", recipients often have feelings of deep gratitude and appreciation for the people who quite literally gave themselves to mend and extend their lives, especially if they received their organ from a living donor. Transplantees often incorporate donors into their own identity, feeling that they are allowing that "other" to live on inside them. Many think of themselves as "us", not just "me".

Can we extrapolate this history to BHT? The scale of BHT is far different than any other transplant. A small body part is not being replaced; rather, it is the entire body complete with two of the most emotionally and socially charged body areas (hands and genitals). Everything below the neck would be alien. Individuals may become confused about the relationship between body and identity in ways that do not parallel the experience of organ or face transplantees and that cannot be predicted based on the past experiences of solid organ transplantation or CTA. The percentage of the body that is now from an "other" creates an unprecedented scale of adaptation and could lead patients to reflect on themselves quite differently from other transplantees. Whereas kidney recipients probably do not think much about the past behavior of the donated kidney, previous actions of an entire body may present emotional obstacles that are hard to overcome. Some may feel disgust by the body's past activities. Knowing that the "body" had committed a crime or terminated a pregnancy, for example, may lead recipients who are bothered by crime and abortion to question their level of responsibility for the past behavior of their new bodies and reflect on the degree of ownership they actually have over themselves as united with a socially tainted body (Turner 1992). People uncomfortable with physical intimacy may be hesitant to touch their new bodies, especially the genitals, knowing that their body is, or used to be, someone else and engaged in behaviors that they object to or find offensive. There are many possible scenarios in which body recipients might find themselves subjectively at odds with their new physical selves.

Will patients have a different presence of mind when they are awakened from the post-operative coma? The answer is that they probably will. When these patients look into the mirror, who, or what, will they see? We cannot predict that answer, but we can assume that patients' self-esteem and identity would be disrupted.

Concerns about disturbances in patients' sense of self and identity are indeed real. The potential for psychological distress can be organized into two sub-categories. First, we should consider the purely mental nonmaterial aspects of humanness. This is the content of the mind, that which we think about. Mental "substance", expressed as attitudes and opinions, self-esteem, social identity, and a person's emotional profile, is our consciousness, our reflexivity in practice in which we think of ourselves as objects in the context of our surroundings. Our mind arises through exchanges with other people and society and becomes an object which we recognize through retrospection and reflection. The mind is the substance of what we think about and how we dwell upon ourselves. The mind is the core of our personhood, our humanness.

Many religions contend that the body and human spirit are separate objects, believing that either a body receives a soul or a soul is assigned a body. The soul has an existence of its own separate from the material body, the latter of which serves only to house the soul temporarily as it makes its way through its metaphysical journey. A core tenet of many faiths is that after the physical death of the body, the soul continues to heaven, hell, or into another body via reincarnation. In Hindu cosmology, for example, the body the soul is returned to is based upon its embodied moral character in previous incarnations. In Christianity, the soul is judged on its merits while embodied on Earth and its continuation to another dimension, heaven or hell, is based on that final judgement.

On the other hand, operationalizing the soul as an encultured individual with stable self-identity, cognitive functioning, and memories leads us to think of the soul as a socialized being whose mind operates in concert with its body. It is this definition that sociologists refer to as the "self". What we see as a living and rational individual is the synthesis of the mind and body intertwined with each other to form a functional whole. The body is the material processor of the mind. We can think of the body as representing the scaffolding that supports and processes our thoughts, emotions, and behavior. We have no mind without the body, but can we have a body without a mind? Clearly no. Without the mind, the body is simply a biological machine void of symbolic significance and reference. It is without humanity.

The relationship between the mind and body is not so elementary or platitudinal. We are learning that the brain, the corporeal "playing field" of the mind, is influenced by what goes on elsewhere in the body. If the body outside the head changes, the brain would likely be affected in turn. If the original body of the head recipient is systematically integrated with the brain, then it stands to reason that the brain will have to adjust to its new physiological neighborhood. It is likely that the brain will notice that it is interacting with a new body. How that engagement might present itself in terms of psychological health constitutes an undeterminable risk. Perhaps the new body will facilitate better brain health. Or it could lessen the brain's well-being and cause the patient immense psychological distress.

BHT and Cerebrocentrism

The proponents of BHT have sent mixed messages about psychological risks. They have stressed the need for psychiatric oversight of the BHT process, beginning with pretransplant screening and then continuing for an unspecified time following the surgery. Canavero (2013:S341) states that after the operation, "body image and identity issues will need to be addressed". He goes on to say, "the patient's perception of the allotransplant should continuously be readdressed by the psychiatrists to ensure that positive, but realistic expectations are maintained". In a later article, Ren and Canavero (2017a) express concern that the characteristics of the donated body, namely the body donor's microbiome, will impact psychological well-being. Though they state that the impact of the body's microbiome on psychological states is "unknown", they, somewhat confusingly in the same paragraph, state that the "microbiome is believed to influence the psyche." It is not clear if they accept the position that the body and the mind intersect or that they function separately.

This ambivalence towards how the new body might affect psychological functioning of the recipient brain is compounded by Canavero and Ren's proposed medical and pseudo-medical remedies. For example, should the microbiome be proven to affect the psyche, BHT proponents recommend a pre-surgical fecal transplant from the donor body to the head's body (Ren and Canavero 2017a). To facilitate psychological adjustment to the transplanted body, subjecting patients to immersive virtual reality and hypnosis before and after surgery has been proposed (Ren and Canavero 2017a).

Despite this concern for patients' psychological welfare, the proponents of BHT have by and large trivialized the possibility that receiving a new body will impact the recipient head's personality, social identity, and self-concept. There is a belief among BHT supporters that the "self" is "highly plastic and easily manipulatable" (Ren and Canavero 2017a: 202) and that the self is an "illusion" (Canavero 2015a) though the self has continuity and will appear as unchanged after the transplant. In one interview, Ren stated, "The person is the brain not the body. The body is just an organ" (Parry 2016).

Advocates of BHT are not alone in this proposition. They are following a proposition known as cerebrocentrism, a position that the self, or the soul as philosophers often say, is enclosed completely within the brain. In this view, the immaterial person, represented by the presence of conscious rationality, self-awareness, and memories, goes where the brain goes. Within the realm of cerebrocentristic thought, dual planes of existence are hypothesized. The mind needs a body and obviously the brain is a material substance, but the interaction between them is casual. Their degree of mutual impact is not significant (Eberl 2017). The brain needs a body, but not necessarily any particular body (Eberl 2022). A body is necessary for personhood, but any body is sufficient for an individual's sense of self to continue.

Figure 2.1 shows a rudimentary model of how cerebrocentrists view the relationship between the material and nonmaterial qualities of human existence. Cerebrocentrism locates the brain as isolated from the body and society. This model is essentially linear and imagines the body as the framework, metaphorically the hardware or infrastructure, that creates and contains the mental activity of the brain. Each relationship is unidirectional. Where the brain goes, the person goes. As the brain goes, so goes the mind.

Cerebrocentrism is largely championed by radical biologists and many psychiatrists who believe that our personhood has little existence beyond our brain physiology and the genetic make-up that determines our idiosyncratic brain structure. This perspective is expressed in the mantra "genes are destiny".

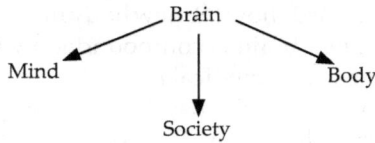

Figure 2.1: Cerebrocentrism.

Is the Brain Really So Isolated?

In contemporary times, cerebrocentrism, has become a minority opinion on the relationship between mind and body. A contrary view holds that the non-material and the material not only intersect but interact. What we see as the person, the immaterial sphere of the mind, is not separate from the material body; rather, it is an expression of the body, and the body is an expression of the mind.

A perspective that counters cerebrocentrism is that social environment, personal psychological attributes, and physiology constantly influence each other (illustrated in Figure 2.2). The elemental formula to understanding mind-body links is not "nature versus nurture" but "nature and nurture". The question here asks: How do the body and the mind interact to form a person's psychosocial self and shape behavior? The brain is where our conscious and unconscious selves are seated. We think with our brain, emotions are processed there, and our behavior is connected to cognitive and emotional interpretations of environmental stimuli and habits of memory, all of which are rooted in brain functioning. Yet the brain does not operate in isolation from the rest of the body or the society in which it is embedded. Boundaries are fluid in these non-linear, reciprocal relationships, making reducing psychosocial characteristics to genes or brain physiology one-dimensional and naïve.

Figure 2.2 depicts a quite different representation of body and mind than cerebrocentrism. Whereas in cerebrocentrism, the brain is separated from mind, body, and environment, in a holistic model, the relationships are reciprocal. The brain both affects and is affected by external stimuli, as is the nonmaterial mind contained within the brain. Mark Cherry succinctly sums up the matter:

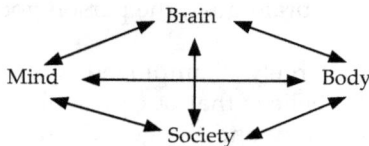

Figure 2.2: Holistic Reciprocal Model.

The self and body can be conceptually distinguished one from another. However, self and body cannot always be physically separated from each other (2022: 244).

These ideas are not new. In the West, theories about the union of body and mind date to the ancient Greeks who recognized that mental and physical health were interrelated and that harmony between the mind and body was the key to well-being (Kleisiaris et al. 2014). More recently, Guiliano Mori[5] reminds us of American psychologist and pragmatic theorist John Dewey, who, writing in the first half of the 20th Century, contended that arguments positioning matter, life, and mind as different and separate entities or essences constituted a flawed philosophy and are packed with errors (Dewey 1929 cited in Mori 2016). Later, Antonio Damásio posited that the brain and mind form an indissociable organism that is integrated by systems of biochemical neural circuits (Damásio 1994 cited in Mori 2016).

What is often missing in biological analyses of the self is the influence of macro forces on body and mind. The mind-body relationship does not engage in a socio-cultural vacuum. Society plays a major role in how our mind and body interconnect. Our sex, race and ethnicity, age, and physical abilities all have social meanings separate from our own thinking. We interpret and mediate these meanings when we engage in reflexivity, the process of thinking about who we are and how we are connected to our social world. Our markers of social status, abilities, and health affect self-identity and self-efficacy, which in turn impact ability and health. Our bodies are our first interface with our socio-cultural world, which provides the boundaries of how we think and what we think about.

The macro-level socio-cultural environment in which we live exerts a strong influence on what happens to our bodies. How we earn a living and exposure to environmental hazards such as industrial waste are two ways in which the social climate taxes and shapes our bodies, which in turn impacts health and life expectancy. Cultural and social forces affect body adornment (tattooing, piercings), sets the rules of attractiveness,

---

[5] In an editorial intending to promote BHT, Xiao Ping Ren curiously refers to Professor Giuliano Mori as a "professional philosopher" (2016: 259). The professional or amateur status of other authors cited in the paper is not referenced. Suspicious that noting Mori's stature as a professional was somehow telling readers that some philosophers were not "professional"—whatever that might mean—and should not be taken seriously, or worse, that philosophers regardless of their status are not equal partners in scholarly pursuits, I did a quick internet search on Professor Mori and learned that he is indeed a professional philosopher. He is a professor at the University of Milan and previously held a prestigious Andrew Mellow Fellowship at Harvard University. It is unknown why Ren, a professional physician and medical researcher, would single out this one scholar. Readers are left to their own devices to make sense of this language.

influences how many pregnancies women have, who gets health care, and determines whose bodies are used for military or criminal violence.

Examples of society, body, and mind connections are countless. Some are quite simple and commonly known. For instance, we are "not ourselves" when we are ill. We are often reminded not to make major life decisions when we feel poorly, and sometimes illness causes our moods to be so unpleasant that our family and friends avoid us until we get well. Stress stemming from external social relationships is another instance. Chronic stress can impact behavior and cognition to the point that we adopt conscious or unconscious coping devices or self-defeating behaviors. Stress can also make our bodies ill. Even our genes are not free from thoughts and behaviors. As each year passes, we learn more about how our behavior and the social environment can change the way genes are expressed. Epigenetics teaches us that behavior and environment do not alter our DNA profiles but change the way our bodies interpret a DNA sequence (CDC 2022) and contributes to explaining a number of conditions ranging from some cancers and autoimmune disorders to psychiatric problems. The list of mind-body connections goes on and on and confirms Jocelyn Downey's precept that "the mind does not emerge in isolation" (2018).

New research shows that the impact of body systems on personality and mental health runs deeper than we initially believed, and three examples are particularly pertinent to BHT. One centers on researchers' discoveries that the immune system impacts an individual's psychological profile. The second is the relationship between gut-biome and brain health, and lastly, the endocrine system and mental well-being are now known to be connected (Downey 2018).

The influence of immunosuppression on mental health lurks in the minds of some researchers who have considered the BHT blueprint. That question ponders how such a severe disruption of the immune system of the new integrated body, that is the body conjoined with its new head, will impact the mind of the survivor (Downey 2018). Evidence suggests that the central nervous system (CNS) wields considerable control over the immune system and, reciprocally, the immune system affects the CNS (Sundman and Olofssen 2014). Of special interest are professional immune cells, those specialized immune cells that present an antigen to a T-cell. T-cells are important parts of our immune response and develop in bone marrow and fight infection and possibly cancer. Professional immune cells can respond to signals from dopamine, acetylcholine, and serotonin, key neurotransmitters believed connected to psychological health. These immune cells can reproduce these same neurotransmitters to exert effects on neurons (Marin and Kipnis 2013; Downey 2018).

In addition, Toll-like receptors, which respond to bacterial infections and mediate inflammatory responses, are known to influence the nervous system, brain development, learning, and memory (Downey 2014).

Individuals have their own immunological "self", but it too functions within a context of psychosocial conditions. Michal Morag's research team (1999) designed a study to investigate the psychological effects of a viral infection and how immune activation mediates those effects. The researchers found, not surprisingly, that viral infections (rubella was the test virus) are associated with short- and long-term psychological distress, especially among individuals from low income families. What was more interesting was their finding comparing two groups based on their levels of antibodies. Teens who, at the onset of the study, were seronegative and then infected with rubella after vaccination, reported more problems maintaining attention and engaged in more delinquent behavior in the ten weeks following vaccination. Teens who possessed higher levels of antibodies and were immune to rubella before vaccination demonstrated less psychological distress. Most intriguing was the interaction with socioeconomic status: poorer teenagers were affected more by their levels of antibodies than their more advantaged peers.

The immune system has been identified as associated with a wide range of psychiatric pathologies and biopsychosocial problems (Downey 2018).[6] This list includes bipolar disorders, chronic fatigue syndrome, anorexia, post-traumatic stress disorder, psychosis, depression autism, narcolepsy, schizophrenia, anxiety, suicidal behavior, and sleep disturbance. Clearly the CNS and the immune system are communicating.

Despite the aforementioned uncertainty shown by proponents of BHT that the relationship between gut-biome and neurological and psychological functioning, research has emerged linking healthy gut functioning to normal central nervous system functioning (Clapp et al. 2017). In 2004, Nobuyuki Sudo's team was the first to suggest the presence of the gut-brain axis, a system in which hormones, neurotransmitters, and immunological factors send signals to the brain directly or through autonomic neurons. We are now learning that these systems are bidirectional—the signals go both ways. A healthy gut predicts a healthy central nervous system, and vice versa.

While it is obvious that dysbiosis, which refers to changes in the taxonomic character of the microbial bacterial composition and distribution, and gut inflammation are associated with gastrointestinal disorders, among many other conditions, it is less known that they are also associated with mental illness, particularly depression and anxiety.

---

[6] I direct readers to Jocelyn Downey's superb review of this literature. This list is attributable to her diligence.

Hormones, neurotransmitters, and immunological cell matter released from the gut send signals directly to the brain and the autonomic nervous system, which contributes to the regulation of involuntary physiologic processes including heart rate, blood pressure, respiration, digestion, and sexual arousal (Waxenbaum et al. 2023; Clapp et al. 2017). The microbiome contributes significantly to the interactions between the gut and the CNS, helping to regulate brain chemistry and influence the neuro-endocrine system associated with the stress response, anxiety, depression, and memory (Carabotti et al. 2016).

These findings have led researchers to experiment with probiotics in the treatment of anxiety and depression. Results are promising, and a couple of examples from this line of inquiry establish the point. Probiotics have been shown to lower cortisol (Schmidt et al. 2015) and reduce negative thoughts associated with depressed moods (Steenbergen et al. 2015). Fecal matter transplanted from psychologically healthy individuals has also been shown to alleviate symptoms in some people with psychological distress (Chinna Meyyappan et al. 2020).

The endocrine system, with its elaborate network of hormone-producing glands, not only regulates most every system in our bodies, but also works in close harmony with the brain. In fact, we have known since the 1950s that hormones play an important role in brain development (Leng 2018). Advances in the relatively new field of neuroendocrinology have outlined the functional interplay between the neuro-functioning and the endocrine system and demonstrated that mental health and overall brain health can be affected by vacillations in hormone levels. Hormones are involved in regulating the immune system, metabolism, brain communication, learning and memory, and neurogenesis, the creation of new brain cells.

Within the brain resides the "command center" of the endocrine system—the pituitary gland also known as the "master gland". Situated deep inside the brain, the hypothalamus is the brain's "junction box" that links the endocrine and nervous systems. It regulates the pituitary, which is attached to the hypothalamus and together the two function to make the hormones that affect and protect virtually all facets of health and well-being.

Unknown risks lurk in the deep recesses of the brain if the hypothalamus and pituitary are conjoined with a new "below the neck" endocrine system. Will the brain accept the new system? Is it certain that the neurological system will assimilate or integrate the networks of glands and hormones? What will happen if the immune system of the body rejects the pituitary gland, which is not protected by the blood-brain-barrier?

The questions generated by this part of the BHT experiment not only have implications for physical health, but mental health as well. Anything less than a smooth integration of the host hypothalamus and pituitary with the donor body could disrupt hormone production, causing disturbances in sexual behavior, learning, memory, and overall mood. Hormone dysregulation could lead to and worsen mental health conditions and when combined with stress, which is surely going to be present in BHT, places patients at high risk of clinical levels of depression of anxiety. Stress increases the production of the hormone cortisol and lowers the body's stock of serotonin, the so-called "feel good" neurotransmitter.

There are many ways in which disruptions in endocrinological functions can affect mental health. One example is that too much and too little thyroid activity usually produces depressive and anxious moods. Another is that changes in sex hormones affect cognition and behavior as well as emotions. The dysregulation within the endocrine system is directly associated with a host of psychological consequences including depression and anxiety, confusion, difficulty concentrating, memory loss, and mood swings. Mental distress can also be indirect consequences of physical changes symptoms of endocrine disruptions such as fatigue, insomnia, muscle aches and weakness, heart rate changes, intestinal distress, hair loss, dry skin, changes in weight, temperature intolerance, infertility, and irregular menstrual cycles among women, among others. It's a long list of potential bodily changes that could impact the mind if the endocrine system does not right itself after a BHT.

All this is meant to suggest that a BHT survivor also risks experiencing psychiatric distress from disrupting the spinal cord and its interaction with the immune system. The drastic steps needed to prevent rejection may further interfere with the links between body and brain health that could manifest in behavioral, emotional, and cognitive disturbances.

There is little question that our bodies are essential to who we are as a person and play a causal role in our uniquely expressed humanness (Eberl 2017).

Conclusion

Body-head transplantation is a life-saving surgical proposal rich in complexity but rife with technical uncertainties. Many steps of the surgery can either fail or produce undesirable outcomes, many of which are so severe that the ethicality of the proposal is under suspicion.

BHT, therefore, constitutes a manufactured risk because it has not yet satisfied the standards of safety expected in modern surgery. The chances of not surviving are high, but even if patients were to live, we cannot say for sure what their quality of life would be. There are too many ways the surgery could produce unwanted results. The key ethical disconnect is that some of the advocates in favor of BHT have promoted the surgery as if these risks were minimal, rather than probable. Many have not been adequality researched, and others have been minimized or discounted or even ignored.

It is possible that had Mr. Spiridonov not come forward as a volunteer and made himself available to the media, BHT may have disappeared as merely a bad idea that had a brief moment in the limelight. But having a living and willing patient made the potential of BHT all too real, and perhaps triggered the medical ethics community's investigation and analysis and general resistance.

In Chapter 3, we will review the ethical conflicts that have emerged from this flow of events and how BHT has been perceived and understood by those who have considered its ethical implications.

References

alekseev.biz. January 31, 2022. How does Valery Spiridonov, who wanted a head transplant, live? Alekseev.Biz. alekseev.biz/news/how-does-valery-spiridonov-who-wanted-a-head-transplant-live/

Barker, J.H., A. Furr, J.P. Barret and M.A. Hardy. 2018. Are we ready for a human head transplant? The obstacles that must be overcome. Current Transplantation Reports, 5(2): 189–198. doi.org/10.1007/s40472-018-0196-7

Beck, U. 1992. Risk Society: Towards a New Modernity. Sage, London.

Bonvillian, C. 30 August 2016. Who is Valery Spiridonov? 5 things to know about Russian volunteer for first human head transplant. The Atlanta Constitution. ajc.com/news/health-med-fit-science/who-valery-spiridonov-things-know-about-russian-volunteer-for-first-human-head-transplant/YxYnYdj1oXQz6yquVXPdhK/ (accessed 6 October 2022)

Borgens, R.B., A.R. Blight and M.E. McGinnis. 1990. Functional recovery after spinal cord hemisection in guinea pigs: The effects of applied electric fields. The Journal of Comparative Neurology, 296(4): 634–653. doi.org/10.1002/cne.902960409

Borgens, R.B., J.P. Toombs, A.R. Blight, M.E. McGinnis, M.S. Bauer, W.R. Widmer et al. 1993. Effects of applied electric fields on clinical cases of complete paraplegia in dogs. Restorative Neurology and Neuroscience, 5(5,6): 305–322. doi.org/10.3233/RNN-1993-55601

Borgens, R.B., A.R. Blight and M.E. McGinnis. 1990. Functional recovery after spinal cord hemisection in guinea pigs: The effects of applied electric fields. The Journal of Comparative Neurology, 296(4): 634–653. doi.org/10.1002/cne.902960409

Borgens, R.B., J.P. Toombs, A.R. Blight, M.E. McGinnis, M.S. Bauer, W.R. Widmer et al. 1993. Effects of applied electric fields on clinical cases of complete paraplegia in dogs. Restorative Neurology and Neuroscience, 5(5,6): 305–322. doi.org/10.3233/RNN-1993-55601

Brissot, R., P. Gallien, M.P. Le Bot, A. Beaubras, D. Laisné, J. Beillot et al. 2000. Clinical experience with functional electrical stimulation-assisted gait with parastep in spinal cord–injured patients. Spine, 25(4): 501–508.

Canavero, S. 2013. HEAVEN: The head anastomosis venture project outline for the first human head transplantation with spinal linkage (GEMINI). Surgical Neurology International, 4(Suppl 1): S335–S342. doi.org/10.4103/2152-7806.113444

Canavero, S. 2015b. The "Gemini" spinal cord fusion protocol: Reloaded. Surgical Neurology International, 6(1): 18. doi.org/10.4103/2152-7806.150674

Canavero, S. 10 April 2022a. World's 1st human head transplantation—Dr Sergio Canavero—Neurosurgeon [YouTube Channel]. youtube.com/watch?v=KY_rtubs6Lc (accessed 23 October 2022)

Canavero, S. and V. Bonicalzi. 2016. Central pain following cord severance for cephalosomatic anastomosis. CNS Neuroscience & Therapeutics, 22(4): 271–274. doi.org/10.1111/cns.12527

Canavero, S., X.P. Ren, C.Y. Kim and E. Rosati. 2016a. Neurologic foundations of spinal cord fusion (GEMINI). Surgery, 160(1): 11–19. org/10.1016/j.surg.2016.01.027

Capogrosso, M., T. Milekovic, D. Borton, F. Wagner, E.M. Moraud, J.B. Mignardot et al. 2016. A brain–spine interface alleviating gait deficits after spinal cord injury in primates. Nature, 539(7628): 284–288. doi.org/10.1038/nature20118

Carabotti, M., A. Scirocco, M.A. Maselli and C. Severi. 2015. The gut-brain axis: Interactions between enteric microbiota, central and enteric nervous systems. Annals of Gastroenterology, 28(2): 203–209.

Centers for Disease Control and Prevention. 2022b. What is epigenetics? Office of Science (OS), Office of Genomics and Precision Public Health. cdc.gov/genomics/disease/epigenetics.htm

Dewey, J. 1929. Experience and Nature. George Allen & Unwin, London.

Chang, G. and B. Pomahac. 2013. Psychosocial changes 6 months after face transplantation. Psychosomatics, 54(4): 367–371.

Cherry, M.J. 2022. What happens if the brain goes elsewhere? Reflections on head transplantation and personal embodiment. The Journal of Medicine and Philosophy: A Forum for Bioethics and Philosophy of Medicine, 47(2): 240–256. doi.org/10.1093/jmp/jhab045.

Chinna Meyyappan, A., E. Forth, C.J. K. Wallace et al. 2020. Effect of fecal microbiota transplant on symptoms of psychiatric disorders: a systematic review. BMC Psychiatry, 20: 299 doi.org/10.1186/s12888-020-02654-5

Clapp, M., N. Aurora, L. Herrera, M. Bhatia, E. Wilen and S. Wakefield. 2017. Gut microbiota's effect on mental health: The gut-brain axis. Clinics and Practice, 7(4): 131–136. doi.org/10.4081/cp.2017.987

Coney, M.G. 1973. Friends Come in Boxes. DAW, London.

Damásio, A.R. 1994. Descarte's error: Emotion, reason, and the human brain. Avon Books, New York.

Douglas, M. 1990. Risk as a forensic resource. Daedalus, Fall: 1–16.

Downey, J. 2018. Head transplantation: The immune system, phantom sensations, and the integrated mind. The New Bioethics, 24(3): 228–239. doi.org/10.1080/20502877.2018.1528808

Eberl, J.T. 2017. Whose head, which body? AJOB Neuroscience, 8(4): 221–223. doi.org/10.1080/21507740.2017.1392380

Emmady, P.D. and J. Bodle. 2022. Werdnig Hoffmann Disease. In StatPearls [Internet]. StatPearls Publishing. pubmed.ncbi.nlm.nih.gov/32644359/

Furr, A. 2022. The Sociology of Mental Health and Illness. Sage, Los Angeles.

Furr, A., M.A. Hardy, J.P. Barret and J.H. Barker. 2017. Surgical, ethical, and psychosocial considerations in human head transplantation. International Journal of Surgery, 41: 190–195. doi.org/10.1016/j.ijsu.2017.01.077

Gerasimenko, Y.P., D.C. Lu, M. Modaber, S. Zdunowski, P. Gad, D.G. Sayenko et al. 2015. Noninvasive reactivation of motor descending control after paralysis. Journal of Neurotrauma, 32(24): 1968–1980. doi.org/10.1089/neu.2015.4008

Gkasdaris, G. and T. Birbilis. 2019. First human head transplantation: Surgically challenging, ethically controversial and historically tempting—an experimental endeavor or a scientific landmark? Maedica, 14(1): 5–11. doi.org/10.26574/maedica.2019.14.1.5

Gołębiowska1, M. and B. Gołębiowska1. 2018. Head transplant—The newest reports and concerns on planned procedure. World Scientific News, 105: 212–217.

Hardy, M.A., A. Furr, J.P. Barret and J.H. Barker. 2017. The immunologic considerations in human head transplantation. International Journal of Surgery, 41: 196–202. doi.org/10.1016/j.ijsu.2017.01.084

Hjelmagaard, K. 17 November 2017. Italian doctor says world's first human head transplant "imminent." USA Today. Usatoday.com/story/news/world/2017/11/17/Italian-doctor-says-worlds-first-human-head-transplant-imminent/847288001 (accessed 25 July 2023)

Iltis, A. 2022. Heads, bodies, brains, and selves: Personal identity and the ethics of whole-body transplantation. The Journal of Medicine and Philosophy: A Forum for Bioethics and Philosophy of Medicine, 47(2): 257–278. doi.org/10.1093/jmp/jhab049

Iske, J., Y. Nian, R. Maenosono, M. Maurer, I.M. Sauer and S.G. Tullius. 2019. Composite tissue allotransplantation: Opportunities and challenges. Cellular & Molecular Immunology, 16(4): 343–349. doi.org/10.1038/s41423-019-0215-3

Keyvan-Fouladi, N., G. Raisman and Y. Li. 2003. Functional repair of the corticospinal tract by delayed transplantation of olfactory ensheathing cells in adult rats. The Journal of Neuroscience, 23(28): 9428–9434. doi.org/10.1523/JNEUROSCI.23-28-09428.2003

Kleisiaris, C.F., C. Sfakianakis and I.V. Papathanasiou. 2014. Health care practices in ancient Greece: The Hippocratic ideal. Journal of Medical Ethics and History of Medicine, 7: 6.

Lamba, N., D. Holsgrove and M.L. Broekman. 2016. The history of head transplantation: A review. Acta Neurochirurgica, 158(12): 2239–2247. doi.org/10.1007/s00701-016-2984-0

Leng, G. 2018. The endocrinology of the brain. Endocrine Connections, 7(12): R275–R285. doi.org/10.1530/EC-18-0367

Lilja-Cyron, A., C.R. Bjarkam and J. Brennum. 2018. Human head transplantation. Ugeskrift for Laeger, 180(31): V01180076.

Liu, F., H. Zhang, K. Zhang, X. Wang, S. Li and Y. Yin. 2014. Rapamycin promotes Schwann cell migration and nerve growth factor secretion. Neural Regeneration Research, 9(6): 602. doi.org/10.4103/1673-5374.130101

Lupton, D. 1993. Risk as moral danger: The social and political functions of risk discourse in public health. International Journal of Health Services, 23(3): 425–435.

McCaffery, M. 1968. Nursing practice theories related to cognition, bodily pain, and man-environment interactions. University of California Print, Oakland.

Marin, I. and J. Kipnis. 2013. Learning and memory … and the immune system. Learning & Memory, 20(10): 601–606. doi.org/10.1101/lm.028357.112

Morag, M., A. Morag, A. Reichenbert, B. Lerer and R. Yirmiya. 1999. Psychological variables as predictors of rubella antibody titers and fatigue—a prospective, double blind study. Journal of Psychiatric Research, 33(5): 389–395. doi. org/10.1016/S0022-3956(99)00010-2

Mori, G. 2016. Head transplants and personal identity: A philosophical and literary urvey. CNS Neuroscience & Therapeutics, 22(4): 275–279. doi.org/10.1111/cns.12534

National Institute on Aging. 2020. How the Aging Brain Affects Thinking. National Institutes of Health. nia.nih.gov/health/how-aging-brain-affects-thinking

Nelson, S.C. June 14, 2015. Head transplant Dr Sergio Canavero is recruiting surgeons for $15m operation on Valery Spiridinov. Huffington Post. huffingtonpost.co.uk/2015/06/15/head-transplant-dr-sergio-canavero-recruiting-surgeons-for-15m-operation-valery-spiridinov_n_7583394.html

Nizzi, M.C., S. Tasigiorgos, M. Turk, C. Moroni, E. Bueno and B. Pomahac. 2017. Psychological outcomes in face transplant recipients: A literature review. Current Surgery Reports, 5(10): 26. doi.org/10.1007/s40137-017-0189-y

Parker, J.C. 2022. Head transplantation and immortality: When is life worth living forever? The Journal of Medicine and Philosophy: A Forum for Bioethics and Philosophy of Medicine, 47(2): 279–292. doi.org/10.1093/jmp/jhab044

Parry, S. March 12, 2016. After gruesome surgery on monkeys, mice and corpses China's Frankenstein reveals: I may do human head transplant next year. The Daily Mail. dailymail.co.uk/health/article-3489595/After-gruesome-surgery-monkeys-mice-corpses-China-s-Frankenstein-reveals-human-head-transplant-year.html

Peters, R. 2006. Ageing and the brain. Postgraduate Medical Journal, 82(964): 84–88. doi.org/10.1136/pgmj.2005.036665

Ren, S., Z. Liu, C.Y. Kim, K. Fu, Q. Wu, L.T. Hou et al. 2019. Reconstruction of the spinal cord of spinal transected dogs with polyethylene glycol. Surgical Neurology International, 10, 50. doi.org/10.25259/SNI-73-2019

Ren, X.P. and S. Canavero. 2017a. HEAVEN in the making: Between the rock (the academe) and a hard case (a head transplant). AJOB Neuroscience, 8(4): 200–205. doi.org/10.1080/21507740.2017.1392372

Ren, X.P., C.Y. Kim and S. Canavero. 2019. Bridging the gap: Spinal cord fusion as a treatment of chronic spinal cord injury. Surgical Neurology International, 10, 51. doi.org/10.25259/SNI-19-2019

Ren, X.P., M. Li, X. Zhao, Z. Liu, S. Ren, Y. Zhang et al. 2017. First cephalosomatic anastomosis in a human model. Surgical Neurology International, 8(1): 276. doi.org/10.4103/sni.sni_415_17

Ren, X.P., K. Luther, L. Haar, S.L. Wu, Y. Song, J.G. Shan et al. 2014. Concepts, Challenges, and Opportunities in Allo-Head and Body Reconstruction

(AHBR). CNS Neuroscience & Therapeutics, 20(3): 291–293. doi.org/10.1111/cns.12236

Ren, X.P., Y.L. Ye, P.W. Li, Z.L. Shen, K.C. Han and Y. Song. 2015. Head transplantation in mouse model. CNS Neuroscience & Therapeutics, 21(8): 615–618. doi.org/10.1111/cns.12422

Rousseau, M.C., S. Pietra, M. Nadji and T. Billette de Villemeur. 2013. Evaluation of Quality of Life in Complete Locked-In Syndrome Patients. Journal of Palliative Medicine, 16(11): 1455–1458. doi.org/10.1089/jpm.2013.0120

Szalavitz, M. 2002. The brain-immunology axis. Cerebrum: The DANA Forum of Brain Science, 4(4): 13–21.

Scheffler, S. 2016. Death and Afterlife. Oxford University Press, Oxford.

Schillace, B. No Date. Yes, you can transplant a head. medium.com. brandy-schillace.medium.com/yes-you-can-transplant-a-head-163bfb099e5d (accessed 27 December 2022)

Schmidt, K., P.J. Cowen, C.J. Harmer, G. Tzortzis, S. Errington and P.W.J. Burnet. 2015. Prebiotic intake reduces the waking cortisol response and alters emotional bias in healthy volunteers. Psychopharmacology, 232(10): 1793–1801. doi.org/10.1007/s00213-014-3810-0

Spitzer, T.R. and D.H. Sachs. 2022. Transplantation tolerance through hematopoietic chimerism. New England Journal of Medicine, 386(24): 2332–2333. doi.org/10.1056/NEJMe2204651

Steenbergen, L., R. Sellaro, S. Van Hemert, J.A. Bosch and L.S. Colzato. 2015. A randomized controlled trial to test the effect of multispecies probiotics on cognitive reactivity to sad mood. Brain, Behavior, and Immunity, 48: 258–264. doi.org/10.1016/j.bbi.2015.04.003

Sudo, N., Y. Chida, Y. Aiba, J. Sonoda, N. Oyama, X.N. Yu et al. 2004. Postnatal microbial colonization programs the hypothalamic-pituitary-adrenal system for stress response in mice: Commensal microbiota and stress response. The Journal of Physiology, 558(1): 263–275. doi.org/10.1113/jphysiol.2004.063388

Sundman, E. and P.S. Olofsson. 2014. Neural control of the immune system. Advances in Physiology Education, 38(2): 135–139. doi.org/10.1152/advan.00094.2013

Suskin, Z.D. and Giordano, J.J. 2018. Body-to-head transplant; A "caputal" crime? Examining the corpus of ethical and legal issues. Philosophy, Ethics, and Humanities in Medicine, 13(1): 10 doi.org/10.1186/s13010-018-0063-2

Tabakow, P., G. Raisman, W. Fortuna, M. Czyz, J. Huber, J., D. Li et al. 2014. Functional regeneration of supraspinal connections in a patient with transected spinal cord following transplantation of bulbar olfactory ensheathing cells with peripheral nerve bridging. Cell Transplantation, 23(12): 1631–1655. doi.org/10.3727/096368914X685131

Turner, B.S. 1992. Regulating Bodies: Essays in Medical Sociology. Routledge, London.

Van Assche, K. and A. Pascalev. 2018a. Full body transplantation, is it allowed? Transplantation, 102(Supplement 7): S223. doi.org/10.1097/01.tp.0000542886.27707.45

Wang, P., H. Wang, K. Ma, S. Wang, C. Yang, N. Mu et al. 2020. Novel cytokine-loaded PCL-PEG scaffold composites for spinal cord injury repair. RSC Advances, 10(11): 6306–6314. doi.org/10.1039/C9RA10385F

Watt, A. 2015. Head transplants: No longer science fiction but a step closer to reality? European Medical Journal. emjreviews.com/neurology/news/head-transplants-no-longer-science-fiction-but-a-step-closer-to-reality/ (accessed 9 August 2022)

Waxenbaum, J.A., V. Reddy and M. Varacallo. 2023. Anatomy, autonomic nervous system. In StatPearls. StatPearls Publishing. ncbi.nlm.nih.gov/books/NBK539845/

Wideman, T.H., R.R. Edwards, D.M. Walton, M.O. Martel, A. Hudon and D.A. Seminowicz. 2019. The multimodal assessment model of pain: A novel framework for further integrating the subjective pain experience within research and practice. The Clinical Journal of Pain, 35(3): 212–221. doi.org/10.1097/AJP.0000000000000670

Williams, B. 2016. The Makropulos case: Reflections on the tedium of immortality. In D. Benatar (Ed.), Life, Death, and Meaning: Key Philosophical Readings on the Big Questions (3rd ed.). Rowman and Littlefield, Lantham, MD, pp. 315–332.

Wolpe, P.R. 2017. Ahead of our time: Why head transplantation is ethically unsupportable. AJOB Neuroscience, 8(4): 206–210. doi.org/10.1080/21507740.2017.1392386

Zielinski, P. and P. Sokal. 2016. Full spinal cord regeneration after total transection is not possible due to entropy change. Medical Hypotheses, 94: 63–65. doi.org/10.1016/j.mehy.2016.06.022

3

The Moral Challenges of the Body-Head Transplantation

Chapter Summary

The previous chapter reviewed the steps proposed to conduct a body-head transplant and the risks that the surgical plan carries. Chapter 3 considers the bioethical conflicts raised by and attached to those risks.

Moral conflicts are found in virtually every step of the proposal. Whereas some are specific, such as the ethicality of decapitating a patient without verification that spinal cord reconnection can result in complete neurological functioning, others are more general and question the very legality of the idea of exposing a patient to a high risk of death.

Most of the ethical questions raised here have not been addressed by proponents of BHT. In fact, ethical concerns have been largely dismissed. The approach taken here emphasizes a pragmatic contrasting "what is" with "what ought not be" rather than a metaphysical or epistemological approach whose questions are existential about the essence of being and existence.

* * *

Paul Wolpe has become an important voice in the ethical debate over the legitimacy of BHT. In one article (2018), he notes that some scholars believe that BHT is so absurd medically and morally tainted that the whole story should simply be ignored. My interpretation of their counsel to disregard BHT is that of a large scale behavioral modification exercise. Collectively snubbing BHT advocates removes the positive reinforcement they receive from publications and media appearances. Theoretically, the behavior, research on and promotion of BHT, should come to an end. In other words, if we ignore BHT, we will "extinguish" it, to use "B-mod" terminology.

For good reason, Wolpe disagrees with this strategy. BHT is not merely a research program, it is a clinical possibility, and at some point, someone may try to implement the blueprint. It is not a time to retreat from the call to attempt a body-head transplant but precisely the moment to assess and review the tenets of this proposal in order to determine if it is acceptable under our current moral infrastructure. Although many have likened the BHT agenda and its discourse to a "circus", it is a spectacle that could move from imagination to reality in the blink of an eye. Because some believe an attempt to perform a BHT is destined to happen, society needs this discussion prior to that first attempt to determine if it is acceptable and morally right.

Canavero and Ren (2017a), BHT's leading proponents, have listed four main objections of the surgery:

1. that spinal cords have yet to be reconnected,
2. ensuring that patients survive the ischemic period,
3. patients' psychological adaptation, and
4. preventing tissue rejection.

These concerns, as Wolpe (2017) notes, are all technical and not ethical concerns. While these may be important clinical puzzles for physicians to solve, they are only one part of the uneasiness that bioethicists fret over when they think about BHT. Most bioethicists believe that Ren and Canavero have not made a convincing argument that justifies conducting a BHT (Wolpe 2017). In fact, the two surgeons have largely avoided direct bioethical questions and have instead relyied on meaningless but emotionally charged rhetoric such as, "When we first announced the experimental data, the world was blown away by the images of the first transplanted monkey" (Ren and Canavero 2017a). As Wolpe has pointed out, such language is misleading, shortsighted, and insubstantial.

Other inconsistencies are present as well. For example, although Canavero and Ren state that psychological adaptation is one hurdle yet to be resolved, they also state that psychological adjustment is of no particular importance. Moreover, they seem to argue that psychological needs do not matter. They have stated more than once that the human self is an "illusion", "easily manipulated", and "highly plastic" (Ren and Canavero 2017a; Canavero 2015), which suggests that psychological adjustment is nothing to worry about. Such a belief posits an ethical dilemma of both content (adjustment cannot be a problem if psychological character is so effortlessly manipulated) and practice (discounting the personhood of patients by saying their sense of self is not real is both inaccurate and contrary to the humanistic value and integrity of an individual).

Keeping in line with pragmatic ethical assessment, what are raised as ethically challenging aspects of BHT are functions of their broader social and medical contexts and their consequences. Ethical flags are raised when BHT's outcomes have moral implications and where there is inconsistency with established practices. Bioethics is about evaluating change and new technology, and where deviation from established norms occurs and could occur, a moral question is raised.

Ethical issues abound with BHT. As we saw in Chapter 2 with a review of the risks involved in this procedure, the risks of BHT are uncertain, and each question of risk begats a torrent of pragmatic ethical problems. The purpose of this chapter is to discuss the ethical shortcomings of BHT, focusing more attention to the possible impact of the surgery on psychological welfare and identity, and the how the failure of mind and body integration could potentially cause devasting psychological harm.

Informed Consent

One of the primary stumbling blocks in making body-head transplantation ethically sound is whether or not a patient, personified as the head donor, can make a truly informed decision and consent to the surgery. Informed consent in a democratic society rests on three key concepts: information, understanding, and voluntariness. Critics, however, have raised fears that the primary proponents of BHT lack the compacity, and even the will according to some, to fulfill these three mandates. In all sincerity, it is a mystery why proponents of BHT have promoted this risky venture given their inability to satisfy this basic bioethical requirement. At this point, consenting to a BHT constitutes little more than agreeing to physician-assisted suicide, which is illegal in most countries and 39 U.S. states.

Physicians have the responsibility to deliver all the information patients need to make the decision to consent to a surgery. The information must be delivered in appropriate language and consistent with patients' reading and comprehension levels. Complete transparency is necessary to ensure that patients have all the information they need to act in their own best interests.

To make their decisions, patients must know what will happen to them, and they must be able to understand that information (Corrigan 2003). This information is more than just procedural steps and intended benefits, but also the likelihood of all negative outcomes and risks. At this point in the development of the technology and expertise needed to make BHT happen, neither of these is possible. Information is incomplete. There are doubts that the spinal cords will reconnect so that patients could regain complete control of both gross and fine motor skills, not to mention the ability to breathe. There are no resolutions on pain and pain management that can be stated without qualification in an informed consent form. Because there are no comparable surgeries to draw inferences and animal research is either not generalizable or insufficient, brain and body integration has not been fully researched and patients' psychological recovery cannot be specified or even extrapolated from other operations. All of these, plus many others, are significant shortcomings in the BHT protocols that must be explained and accounted for in an informed consent document.

BHT is a complex proposal. In truth, it may be the most complicated medical design ever conceived. Any patient, even a physician, would have difficulty understanding all the layers of physiological and psychological disruptions that BHT could cause. Patients would need far more than a lay knowledge of immunology, orthopedics, neurology, and psychiatry to fully grasp what they were agreeing to undergo. Patients might fill in the gaps of their understanding with trust in their physicians, but that strategy comes with risks if there is no open and fully transparent doctor-patient relationship.

Those who criticize BHT procedures note that the designers of those procedures have neither fully disclosed nor addressed all the risks that BHT entails. These critics point to Canavero's multiple and demonstrative statements to the media and in journals that the surgery "works" and that patients will be up and ambulatory in short order with no significant psychological changes.[1] For the surgery's advocates, success is a *fait accompli*, the outcome has been decided in advance of absolute proof, which has the unfortunate consequence of leading patients to trust the surgery will save them and make them better than ever. But the proclamations of the surgery's advocates are made without evidence or sound theoretical justification, yet taken *in toto* and *prima facie*, they could sway patients to give consent not fully informed of the risks. Lacking confirmation of the process, and presenting the process as safe and sound, would negate any informed consent statement presented to a hospital considering supporting a body to head transplant and perhaps expose the hospital to liability risks. Publicly saying the surgery will work without presenting

---

[1] Canavero and Ren often use declaratory and somewhat inflammatory wording that gives the impression that BHT is ready for a human trial. While such language is common in their media presentations, editorialized phrasing is avoided in academic and technical writing. Nonetheless, subjective jargon is common among their papers. Here are a few examples.

- "Houston, GEMIMI has landed: Spinal cord fusion achieved".
- "In June 2013, the world was taken by storm by the announcement that a full head (or body) transplant was possible".

From Canavero and Ren, *Surgical Neurological International*, 2016b.

- "The debate on HEAVEN is raging. However, the reaction from media and academe up to now has been often hysterical and frankly ridiculous".

From Ren and Canavero, *American Journal of Bioethics Neuroscience*, 2017.

- "Hence, we thank this journal [*Surgical Neurological International*] and its editor in allowing us to make its readership apprised of this simple fact: that academic arrogance once again is stifling scientific innovation".

From Ren and Canavero *Surgical Neurological International*, 2016.

written evidence to that effect is a reckless and negligent practice and therefore unethical by democratic standards.

Patients who are considering their chances with BHT are doing so in mental state that compromises their true voluntariness. These patients are suffering a terminal disease and dying and likely in a state of despair if they are considering this desperate last chance for survival. As understandable as wanting to hold onto life may be, anxiety over an imminent death is not sufficient grounds for undergoing a medical procedure that may allow one to live but in a state that could be worse than before the surgery. For survivors, severe central nerve pain, quadriplegia, and psychiatric disturbances are more than possibilities—they are likely. Such conditions would be, as many critics have contended, worse than death. They would be torture.

Desperation compromises the voluntariness required to sign an informed consent statement. It is true that no informed consent is provided in a vacuum as patients and research subjects have motivations or characteristics that separate them from those who take few risks and do not participate. Nonetheless, despair and fear diminish autonomy to the degree that at some point, the scales are tipped so that a patient may agree to anything to try to save their life.

In essence the question about informed consent is to ask if BHT is in the best interest of patients. Certainly, Canavero and Ren believe it is, and patients may deem it in their interests as well (Gelfand 2017). That does not mean that an IRB, institutional leadership, and the law see it that way. A surgery that cannot show evidence of success and could be dangerous for a surviving patient is one that only offers speculation and hope. A third party, an IRB for instance, is a necessary agent of social control to prevent such a surgery from happening. Speculative hope is not grounds for arguing patients' best interest, especially if the outcomes may be worse than the patient's current condition (Gelfand 2017). As Zaev Suskin and James Giordano (2018) have stated, *caveat emptor* is an insufficient basis for informed consent. The patient is not and should not be individually responsible for assessing the veracity and appropriateness of such an extreme experiment.

Ana Ilitis (2022) raises one last intriguing thought about informed consent. If the survivor of a body-head transplant is indeed a third person distinct from the individuality and personage of either of the donors, does that "new" person have rights separate and apart from the donors? Since the new personage does not exist before the surgery, that person cannot give consent. This is not a trivial matter because it is that person, not the humanness of the intact individual donors, who will experience the consequences, which as we have stated, are likely to cause severe physical and mental impairment. It is that person who will experience the hardships of recovery, the pain, and the mental anguish that may follow should they survive.

Giving consent for body-head transplantation violates the standards of the Belmont Report and the other important and generally agreed upon declarations of ethical standards. A person cannot consent to their own death unless that death is temporary and necessary to accomplish the stated goal of the surgery (as is the case with many heart surgeries). Although BHT in that sense is theoretically consistent with established medical practice, there is no reasonably sound theoretical or clinical paradigm in which decapitation is recoverable and will result in improved health. The risks and burdens to patients (and research subjects) must be proportional to the benefits. Agreed, the benefit is the restoration of life for dying patients, but this consent occurs without substantial evidence of viability of the procedure and compromised voluntariness. Under the rubrics of a democratic distribution of burden and risk and transparency, acquiring and giving informed consent for BHT are dubious at best and reckless endangerment at worst.

Legal Uncertainties and Familial Confusions

The existential and phenomenological experiences a BHT survivor will encounter are just one aspect of the psychosocial environment that will change after surgery. As we have already discussed, Western psychology and metaphysics dictates that personhood belongs to the head that contains memories and the ability for reflexive thought. Will the law see it that way? It seems simple on the surface. If the law acted consistently with Western sensibilities, the phenomenological experience of the head, which presumably the body can no longer experience in its original state, determines the legal standing of the "new person" as that of the head. The rights of body donors end at the time of their absolute death, which will occur once they are decapitated. The head donor, therefore, continues as "the person" but there are many legal conditions and processes that must be sorted out before we can say this problem is solved. The torso and limbs account for 80 per cent of our body mass. Are there situations in which that majority has priority over the head and its "command center" functions? A number of scenarios make it difficult to figure out who would be alive after the surgery.

Having the head of one person and the body of another poses legal questions of identity, familial accountability, and responsibility for the actions of the body and head prior to the surgery. In addition, a number of critics have questioned whether BHT, as it stands now at least, is even lawful at this time. Would surgeons be subject to criminal charges if they attempted a body-head transplant and the patient died or was left in worse condition than before the surgery? The legality of the experiment is subject to debate given that it involves the unusual termination of one life and potentially a second. Many legal issues must be resolved before the surgery is conducted. This section will consider the legal ambiguities and risks for patients and surgeons.

Family and Inheritance

Establishing legal identity of the "new person" before a body to head transplant procedure takes place is essential (Suskin and Giordano 2018). Failing to set legal boundaries of identity could lead to confusing family relations, bitter and depressed feelings, and even litigation. To avoid these problems several legal decisions must be made to enhance the ethical practicality of BHT. It would be contractually and interpersonally irresponsible to avoid anticipating how the law and feelings of entitlement might interject themselves in the course of a BHT. Therefore, legally binding and emotionally satisfying agreements between the families of the body and head donors and the head donors themselves must be reached prior to the surgery.

Marital Status

Marital status is one piece of the legal/identity puzzle. If both body and head were married, do they remain legally wedded to their spouses after the surgery? Is a head donor missing 80 percent of its body mass still married to a spouse who may not fully recognize their partner? Would the spouse of the body recognize their partner's body, and, if yes, how should they react? Does the now widowed spouse have any entitlements such as income from investments, pensions, Social Security, and inheritance, earned by their late partner's body after the transplant, especially if the surviving "new person" engaged in manual labor or worked with their hands to create that wealth? After all, 80 percent of their spouse remains alive. Is the body recipient responsible to or accountable for the body's partner?

Parental Status

Many of the same questions raised about spouses also pertain to children born to the two donors before the surgery. Do the children of the body that created them have legal entitlements and ties to their parent whose body remains alive? Should they inherit wealth earned by the new body, though that body now belongs to another person? Body donors, and thus the "new person", are genetically close to their children. Does that carry rights and privileges?

Deciding who is alive before the surgery becomes a matter of some urgency when thinking about the many directions identity can be defined for survivors (Ilitis 2022). These questions may seem rather trivial, since, in Western thought, the survivor will be defined as the personage represented by the head. But how the families will feel is another unknown aspect in the course of conducting BHT, and relatives may turn to courts of law to settle their claims.

Family continuity represents more than genealogical curiosity and feeling part of the historical flow of the generations. It also produces and passes along intergenerational wealth, medical history, and emotional attachments. In some cultures, attachments to ancestors and descendants also have religious significance. Family members of the body donor may feel they have certain rights because their kin's DNA continues to live and may assert that DNA is the definition of who is alive after the transplant.

Whether or not the "new person" can truly reproduce is at one time a medical, phenomenological, and legal poser. Is it ethical for the head donor to use the body donor's reproductive organs to have a child when the donor is deceased (Cuoco 2016)? If BHT patients were to have children, questions over parentage are possible if not likely. The parents of the body donor, who would share DNA with any offspring, may want visitation privileges to children they may see as their grandchildren. Joshua Cuoco also worries about how children may react to learning that their "natural" parent died before they were conceived.

A last concern about reproduction is age differences between the head and body, especially if the head donor is of fertility age but receives a body that is not. This dissonance could be stress inducing for some people and have an impact on identity and aspirations to have children and meet life goals.

In sum, marital, fertility, and parenting deviations from normative practices constitute a serious ethical dilemma to body-head transplantation (Mirkes 2018). While legal matters of who is married to whom and parentage could be contracted prior to the surgery, emotional connections might not be so easily negotiated. Signing a release form does not guarantee the waiver of any legal claims a spouse, parent, or child may make against the new person or emotional bonds they may have to the still-living body.

In these legally and emotionally sticky situations concerning family and identification, the ethical solutions will not be worked out by ethicists and IRBs, but by the patients themselves and the people in their own social *milieu*. What is right and wrong would be negotiated as the events occur, setting precedents for future transplants. If mediators fail to bring the parties together in mutually satisfying agreements, the law may become involved in settling family disputes and in some cases, juries may be forced to decide what is ethical. Resolutions of these gray ethical areas would perhaps be the least authoritarian or patriarchal in that they would be decided or negotiated with minimal institutional intrusion and control.

Nonetheless, they should be resolved in advance of any body and head transplant.

Show Me Your Papers!

Family matters may not involve the state if the families and the head donor contractually agree in advance of the surgery as to who is related to whom and resolve matters of inheritance, property ownership, child visitation, and so forth. These issues will, and could, remain civil matters and not necessarily become subject to statutory regulations. Where the state might intervene, however, is identity verification.

Imagine this scenario. A BHT survivor wants to serve as a volunteer tutor for a nonprofit agency dedicated to improving children's literacy levels and reading skills. To be accepted into the program, volunteers must undergo a background security check, which includes being fingerprinted at the local police station. The former patient forgets to tell police that their fingerprints are not their own (assuming they identify as the person "in the head" who had different prints, that is). After scanning the prints, the police computer sounds an alarm that the fingerprints match those of a person who has committed a series of violent crimes including armed robbery and assault for which, despite having a DNA sample, no suspect had been identified. The BHT patient is immediately arrested and photographed and a DNA sample is taken and matched to that collected at the crime scene . The police, thinking they have apprehended a social menace and fugitive from justice, charge the BHT patient with crimes that the head has no memory of committing or possesses the propensity to commit. What happens now?

For all intents and purposes, the police have no reason to believe they have arrested the wrong person. Fingerprints and DNA do not lie. They are indisputable evidence in a court of law. Herein lies another ethical soft spot: how do "new persons" identify themselves legally? It is unlikely that a BHT patient would be held criminally accountable for the misdeeds of the body prior to its brain death, but what time, resources, and emotional stress must be expended to extract themselves from legal harm in situations like this? Will BHT patients have to register with the state? That potentiality raises more ethical and civil rights questions than it resolves.

There are many other potentially conflicting ID-related situations. Bank accounts, passports, and citizenship are just some of the ways in which the state or an institution requires formal identity-proving documentation that could be compromised by having the DNA, fingerprints, and biometric characteristics of two people (Cherry 2022). Height, which is noted on U.S. drivers' licenses, may change, causing a police officer to question the validity of the license in a routine traffic stop. More and more, Americans are required to show identification to do routine activities such as purchasing medicines at drug stores, receiving medical care at doctors' offices, getting through airport security, or simply turning on their computers. Any discrepancies in IDs could disrupt these routines and add suspicions or trigger time consuming and aggravating hassles. In some cases, an attorney may be required to remedy these situations, adding expenses the patient did not expect to incur.

Could BHT Surgeons' Face Criminal Culpability?

There is one additional legal matter to introduce and that is whether or not BHT is legal. Of course, laws vary from one country to another, but BHT most certainly presents as a legal "gray area" for surgeons' and hospitals' legal teams and the law to decipher. In two countries where developmental research on BHT has been conducted, China and South Korea, the legality of the procedure has already been questioned (Lei and Qui 2020; Kuk and Ryu 2019).

Insurance companies' willingness to extend coverage to experimental BHT would play a role in the legal interpretation of BHT should the surgery fail and the head donor die or become incapacitated. If insurance companies compensated patients and their families for physicians' malpractice in a failed BHT, perhaps there would be less interest in pursuing the legal question of intentional recklessness. Prosecutors might see the surgery as more legitimate if insurance covered it and then interpreted it as a civil matter of possible malpractice rather than criminal negligence. Physicians acting with or without liability insurance face legal exposure in any circumstance, and in the case of BHT, they may be forced to defend themselves against more than charges of civil malpractice. Police may accuse them of committing homicide.

Even if their insurance were to give surgeons financial protections from charges of malpractice, it would not necessarily protect them from families and police defining their experiment, given all its risks, as an act of intentional recklessness. Can physicians and hospitals acquire exemptions from legal liabilities before conducting the surgery? Can they, in a sense, secure a pardon in advance of a possible crime, especially if that crime is homicide? Probably not in a democratic state.

Let's review how police and prosecutors might see body-head transplantation and how this surgery could be interpreted as murder.

The body donor's life will certainly be terminated in a novel way (decapitation), but in a manner consistent with other brain-dead patients who agreed, or their families consented, to donate their organs. Prior to the surgery, the body donor was already pronounced brain-dead but does their final death come as an ordinary organ donor, "ordinary" in the sense of established organ gifting practices, or as a result of a surgery in which evidence of success was not sound? Should the experiment fail, the donor's body would remain unused and wasted in the sense that its organs would no longer be useful for another transplant. The intentions of the organ bequest would be unfulfilled. Ethical and legal questions could be raised at this point, but let's be content with the notion that the death of the body will not generate legal ambiguity.

The head donor's death or permanent disability, however, could be a different matter. It is conceivable that prosecutors could investigate the physicians under tort law for unlawful termination of life because the BHT procedures are unproven, expectations are murky, and informed consent was granted out of desperation. For many reasons, BHT is perhaps the riskiest surgery imaginable and involves consenting to the unprecedented surgical step of decapitation (Van Assche and Pascalev 2018a). Is this a *de facto* consent to homicide, death by surgeon?

Some, such as Kristof Van Assche and Assya Pascalev (2018b), have suggested that surgeons performing a body-head transplant may face charges of reckless homicide should the head donor die as a result of the operation. Tort law would assuredly come into play if the patient had not given consent, but Van Assche and Pascalev believe that criminal proceedings could begin even with consent should something go wrong. A prosecutor could conceivably allege criminal liability against a physician if the patient died and if the conduct that led to the death was deemed "outrageous" or reckless. At the center of the debate would likely be whether or not the physician conducted a surgery that had little probability for success, particularly if the patient were told it would be successful. Canavero has stated publicly that the chance of success is 90 percent (Osborne 2017b), a figure most agree, is highly optimistic and unreasonably ambitious for an unproven experiment. In today's climate of litigation and institutional distrust, decapitation and the probability of paralysis and chronic pain, which are the likely paths of BHT survivors, physicians who perform this experiment within the current boundaries of science would risk not only their license to practice medicine but their freedom as well. A judge and jury may see BHT as a legitimate and heroic attempt to save a life, but they could easily interpret BHT as an experiment that went wrong because the patient was not fully informed of the risk, the medical preparation was shortsighted, or informed consent was questionable.

Proponents could argue that patients were terminally ill prior to consenting to BHT. The surgery, if it went wrong, would do little more than hasten the inevitable and quite possibly give them a better death than that which they may have faced with their disease. Afterall, they might say, patients who die in surgery typically do so painlessly in their sleep. Knowing that death awaits them, patients have the chance to sort out their affairs, say their goodbyes, and pass quietly and peacefully under sedation. Proponents might also argue that patients' have the right to try an experiment that might save their lives. Does this make it right? Not necessarily.

When patients have been informed of risks and given consent, the point at which experimental surgery crosses the line from legitimacy to criminal is uncertain and subjective. Unless there were sound medical reasons, physicians straying from time-proven surgical procedures that are known to produce desired results would be subject to charges of malpractice and criminal recklessness should the patient die or be left with a permanent disability or health problem. In experimental surgeries, however, the determination of recklessness is more vague.

Surgical advances occur in two ways (Petrini 2013). The most frequent method that improvements occur in surgery is through gradual evolution. A procedure is established and through careful observation and study, the procedure is tweaked little by little to improve success and reduce risk. Second and more rarely, improvements arise from groundbreaking experimentation. In a Kuhnian revolutionary sense, the bold experiment is without history, it is a new paradigm that is radically original and unprecedented. BHT is presented by advocates as such a revolutionary solution to life-saving surgery.

As mentioned earlier, unlike the development of new drugs, there is no regulatory equivalent to the U.S. Food and Drug Administration that approves experimental surgeries or monitors surgical safety in the way that pharmaceuticals are reviewed for safety and effectiveness. In lieu of institutional governance, surgeons and hospitals are largely self-regulating, controlling their behavior through ethics boards and intra-discipline monitoring and licensing. Surgery does not have clinical trials, biostatistical controls, or double-blind experimental designs. Deception and "sham" surgery, or a placebo operation, which would be necessary to apply experimental designs to surgeries, are generally considered unethical[2] (Petrini 2013). In the absence of random clinical trials, experimental surgeries are frequently awarded more leniency in that under some circumstances innovations can be attempted in "real" surgical conditions, not under laboratory and methodologically controlled situations comparable to FDA

---

[2] A sham surgery is a term often used to describe the control group in a surgery experiment. It is equivalent to a placebo. A surgeon anesthetizes a patient, makes an incision, then stitches the incision without doing any actual medical intervention. The goal is to determine if the fake surgery produced any functional change compared to an actual surgery in which a medical remedy was attempted. Most bioethicists and surgeons consider sham surgery as fraudulent and unethical because patients are exposed to the risks of anesthesia and the incision, which could become infected, not heal properly, or cascade to other health problems. Methodologically sham surgery cannot meet the standards of a double blind design—the surgeon will know who has the real surgery and who has the fake. Nonetheless, some contend that placebo surgery is ethical if (1) patients give consent and fully understand the risks involved, and (2) compelling scientific reasons justify the sham is necessary to gain medical treatments where no known intervention exists (Petrini 2013).

protocols for pharmaceuticals. Surgical innovations are often validated in "real time" though the expectation of animal trials is usually expected when attempting to make great leaps in advancements.

Despite the call by BHT advocates that patients have the right to try a procedure that might save their lives, the law may not necessarily agree. The Right to Try law passed in the United States in 2018 stipulates that patients suffering from terminal diseases do have the right to access drugs that are currently in clinical trial if the patients are not part of the study. To receive these as yet unapproved medications, patients must satisfy certain criteria such as having exhausted all approved and standard treatments and are likely to receive benefits that outweigh any risks, among others. Considered a compassionate measure, this law provides patients an opportunity to explore medications that might help them but which they were excluded from trying perhaps simply because they did not reside near the hospital or university conducting the research. These drugs have already been approved to qualify as investigational drugs, so there is an assumption that they have met certain efficacy and safety standards. Right to Try does not apply to surgeries because they are not subject to FDA developmental standards. If surgery were covered formally under Right to Try, however, BHT would still not pass moral muster because of its extreme risks. Therefore, there is no ethical reason to apply "right to try" to body-head transplantation.

What this means for BHT, and most all composite tissue allotransplantations, is that the first patient is also the first experimental case, which may afford some degree of legal protections for surgeons. Surgeons must still secure consent and present evidence of risks and outcomes gained through animal experiments and extrapolation from other comparable surgeries. The legal fate of the surgeons would rest upon that data and how they present the likelihood of success. The data and the inference of success should correspond.

Even if the surgeons had IRB approvals and informed consent, risk remains should a failure occur. Patients and their families could question the evidence of support for BHT, as it currently stands, and argue that physicians were not fully forthcoming in acquiring consent and liability waivers from patients prior to surgery. Without satisfying these extant regulations and receiving the necessary institutional and legal permissions and clearances, however, it is likely that BHT surgeons would indeed be vulnerable to legal and civil charges of criminality and malpractice if the patient were to die or left in a health condition worse than that before the surgery (Van Assche and Pascalev 2018a).

The legality of BHT may rest upon the legal system's definition of the surgery. Is it characterized as a research project or a medical experiment? The differences have important legal implications. Van Assche and Pascalev (2018a) are convinced that under current technological conditions, BHT does not conform to the legal standards and expectations of research. This position is reinforced by the numerous medical and ethical objections levied against BHT. If BHT were defined as a medical intervention, current law could place physicians in jeopardy of facing charges of medical malpractice or even negligent homicide should the patient die (Van Assche and Pascalev 2018). Again, this judgement would be based on the lack of evidence that the intervention will succeed and on questionable consent.

The charge that BHT is homicide is subject to local laws and their interpretation by prosecutors and juries. A case could be made that patients agreed to undergo an experimental surgical protocol with important questions unanswered and a low likelihood of the intended outcomes of restored health, full ambulation, and additional years added to their life expectancy out of desperation to save their lives. In this scenario, it could be argued that patients fell victim to a manipulative and unrestrained surgeon acting out of hubris and the desire for attention and glory. Prosecutors could argue beyond malpractice and state that BHT constituted criminal assault because surgeons offered no practical evidence that predicted success, which would be the case with present technology. BHT physicians' may argue that, while their behavior was intended, they did not intend to cause the patients' death, though that outcome is likely. In such circumstances, the physicians' behavior could be interpreted as reckless or negligent, behavior to which patients cannot give consent (Farahany 2016).

Such a scenario has been presented by two philosophers, Ruiping Lei and Renzong Qui, in the context of Chinese law. Lei and Qui (2020) argued that Chinese legal codes could easily interpret BHT as criminal recklessness and unintentional homicide because the physicians had no intent to kill the patient. Since the surgeons' aim would be to save the patient, a patient's death would conflict with that objective, creating a potential legal matter. Lei and Qui state that from a legal viewpoint Chinese homicide is a "subjectively intentional concept". There would be no intent on the part of surgeons to kill their patient, but if death occurred, it could be considered "death caused by negligence". Lei and Qui summarize their point in the following hypothetical case where person A is the person of the head, person B is the body, and person C is the new person, the amalgam of the two donors.

> Although they [the surgeons] may intend to preserve or even improve A's life, if the transplantation succeeds, a new mixed body C might be the result and the original A might be dead; if it were to fail, A would die directly on the operating table. Although A's death is not intended by the surgeon, but her negligence causes A's death.

Parenthetically, Lei and Qui also point out a second legal problem physicians may face if they attempt a BHT in China. They could be charged with abuse of a corpse. Lei and Qui cite Article 302 of the Criminal Law of the People's Republic of China that stipulates that the crime of insulting a deceased body is punishable by imprisonment, detention, or "putting under surveillance" for up to three years. Chinese jurists have argued that decapitating a brain dead body is disrespectful and a violation of human dignity and possibly a violation of the law (Lei and Qui 2020).

A counter argument could be postulated, however. Patients currently make urgent decisions to accept risky treatments to save their lives within the ethical domain of "right-to-try". Cancer patients are routinely invited to participate in various experiments for chancy new treatments that may kill tumors. All remedies that are common today were once novel and experimental, and someone had to go first. Historically, of course, many of these patients pioneers did so without consent and even against their will, but that has not been the norm in Western-oriented societies since ethical reforms occurred after World War II. Patients also routinely agree to surgeries that cause their temporary death. Many heart surgeries, as is obvious in the case of heart transplants and coronary artery bypass grafts, require surgeons to suspend the life of their patients to achieve the desired outcome. Legal precedent for patients to give consent to terminate their own lives, therefore, has been set and is well established as long as the cessation of life is required to achieve the intended outcome of improved health and longevity (Suskin and Giordano 2018).

While on the surface, BHT may seem to fit Nina Farahany's (2016) notion that BHT could be seen as either "active euthanasia" or reckless homicide, the reality is not that clear. How BHT might be interpreted legally is subject to the political mood of the jurisdiction in which it occurred and the cultural orientation and sensibilities of jurors towards life and the medical establishment, not to mention the persuasiveness of attorneys. Should a BHT occur within the confines of present technology, a test case would probably be required to determine the legal standing of surgical procedures that stretch the standards of normative medicine. Such a test case would presume the death of the head donor and the family or police then bringing charges of recklessness and unlawful death. In a legal system such as that of the United States, a jury would decide the fate of the surgeons.

Research Ethics and Scientific Discourse

In the early 1900's, a movement was underway that would transform the very heart and soul of American medical education and practice. The efforts of the Hopkins Circle, the leading voice of the reform movement, which was based at Johns Hopkins University, produced the Flexner

Report in 1910, perhaps the most critical document in American medical history. The report, largely written by Dr. Abraham Flexner based on observations gathered from his tour of North American medical schools, directed physicians not only to master the biology of medicine but to be considerate of the personal narratives and the socio-cultural context of their patients and the conditions they treat. Since then, the narratives surrounding medical science and particular treatments have proven as critical as the science itself in the development and delivery of health care. We can see examples of the intrusion of political discourse into the perceived efficacy and safety of vaccinations and the public response to AIDS. In the past epilepsy was believed caused by demonic possession and the disenfranchisement and stereotyping of people with epilepsy still continues despite our knowledge that the disease is a neurological disorder over which patients have no control. Some conditions, like cancer, extract public sympathy; however, a number of health problems such as leprosy and mental illness are subject to stigmatization or are not accepted by mainstream medicine and public sentiment as legitimate disorders.

The narrative encircling the history of epilepsy is not the only case in which the absence of a verifiable biological foundation subjects the disease to a non-scientific discourse. Numerous diseases have moved from socially and medically questionable to become recognized as legitimate by the mainstream medical institution. People suffering from fibromyalgia were (and sometimes still are) told their chronic pain was psychological and attempts to explain it were dubbed pseudo-science. Now, the condition is recognized as a legitimate disorder though it is not well understood, and that implies fibromyalgia's narrative has changed, redefining it as a real disorder. Electromagnetic hypersensitivity (EHS) is another condition considered illegitimate by medical authorities in some social spheres but not in others. Because studies on EHS reveal evidence that both confirms and denies its existence, the narrative surrounding EHS determines its "reality". These examples show that conditions move into and out of medical definitions. In some cases, this definitional migration is not necessarily due to evidence that supports one position or the other but because the social narrative changed.[3] Homosexuality was once defined as

---

[3] Homosexuality, for instance, lost its medical legitimacy several decades ago when the American Psychiatric Association (APA) voted to demedicalized sexual orientation (Drescher 2015). Relatedly, the APA's definition of mental illness presented in the various editions of the DSM has evolved over the years despite no changes in clinical or neurological evidence. As psychologist Eric Maisel wrote in *Psychology Today* (2013), "the very idea that you can radically change the definition of something without anything in the real world changing and with no new increases in knowledge or understanding is remarkable." The net result of these changes is that the number of mental illnesses has about tripled (Furr 2022).

an illness, and now it is not. Until the early 20th Century, alcoholism was a sin. Then the Freudians redefined it as a character problem, and now is labelled a disease or mental illness. Not that much has changed to better understand the origins of alcoholism, but as society changed so did its narratives.

At this point the sociological axiom derived about a century ago comes into play: if people define situations as real, they are real in their consequences. The famous theorem created by W.I. Thomas and Dorothy Swaine Thomas (1928) means that the interpretation of a situation causes social action and embodies that situation as having a reality of its own. That "reality" is created by a discourse that shapes thought and directs behavior. In other words, how we define something as real, or not real, affects how we respond to it. When individuals present with questionable complaints that are stigmatized, that the social response affects the quality of care they receive. In short, our perception of a health condition is critical in understanding how we treat it.

The general narrative surrounding BHT is that the proposal lacks medical and ethical legitimacy; thus, advocates have not received permission to perform the experiment in a number of countries. Critics contend that BHT will not work medically and that it is morally corrupt, dangerous, and possibly criminal. Media and scholarly presenters, people and organizations that have contributed to the narrative around BHT, have shaped this discourse. The media have used words such as "bombastic", "controversial", and "sensationalist" to describe the proponents of BHT, and members of the academy have largely attempted to discredit BHT. Bioethicist Arthur Caplan of New York University School of Medicine, perhaps the champion narrator of the anti-BHT argument has described BHT as "both rotten scientifically and lousy ethically…one would have to be out of one's mind" to attempt the surgery (2015). A couple of years later, Caplan described the BHT blueprint as "implausible" and that it deserves "not headlines but only contempt and condemnation" (2017). For the New York University professor, head transplantation is truly "fake news" (2017). Other scholars have contributed to the discourse with scholarly arguments against BHT, and most of these are included in the present chapter.

In their attempts to control the narrative, proponents have sought to deliver much of their message via the media. It would appear that the media arena is their preferred method of scientific communication.[4]

---

[4] Canavero continues to rely on the internet to endorse his theories and intentions. In a 2023 interview on Vice.com, Canavero used the platform to promote brain transplantation (Wells 2023). Saying that head transplantation would not solve the problems of an aging head, Canavero discussed his publication in *Surgical Neurology International* (2022) that presented his theory that a brain transplant was indeed possible and developmental research should be supported.

Internet searches on head transplantation and Sergio Canavero reveal too many interviews and presentations to count. Many are well produced with illustrations and models; others are more casual. One of the themes that runs through the advocates' discourse is that BHT is a radical procedure that challenges the tenets of a stodgy medical institution that is resistant to new ideas. Ren and Canavero appear to express more frustration with the academic establishment for rejecting their research agenda than responding to their critics' scientific and ethical objections (DiSilvestro et al. 2017). One example taken from an editorial published in *CNS Neuroscience and Therapeutics* by Xiaoping Ren in 2016 describes that frustration succinctly. Referring to criticisms of BHT as medically and ethically unsound, Ren says:

> Yet there is a point that must be tackled. And that is as follows: Why HEAVEN/AHBR [their code name for head transplantation] met with so much skepticism. Why so much acrimony for a life-saving procedure? The reason is psychological: HEAVEN/AHBR opens the Pandora's Box of medical failures.

Ren continues the point saying:

> Biochemical/Genetic Medicine at large has failed for chronic conditions, insomuch as no quick, definitive cure is available for any condition.... Had medicine succeeded in rooting out the genetic and biochemical causes of chronic degenerative conditions, but also cancer, we would be free—or on the way to—of the major killers that affect mankind. There would be no need for a full body exchange. HEAVEN/AHBR bears brutal testimony to this simple fact.

These thoughts are significant in that they expose the author's irritation at the establishment for having rejected BHT. This narrative has nothing to do with the viability or ethicality of BHT, but rather positions BHT architects as embittered contestants in a failed public relations battle. Saying that BHT is necessary to saves lives because of the failure of traditional western medicine comes across as "ad hoc rescue" logic.

The ad hoc rescue fallacy states that in the event of feeling desperate to be correct, beliefs are held onto despite contrary evidence. In the course of the despair, errors in logic can occur by simply making up excuses as to why the held belief is true and the alternatives are false. In an attempt to make the conflict go away, this new assumption is created ad hoc to give the idea credence and discredit the alternative. The reasons that traditional western medicine has not cured cancer is because cancer is complex and multidimensional and treatments are complicated. Saying "I am right because you have failed" is simply a smoke screen device to sway uninformed opinion to accept the offered "rescue" and divert the narrative to situate BHT in the positive. It is a meaningless argument in a scientific context and a fallacy of logic.

There is further evidence of rhetorical misdirection in the narrative to promote BHT. Wolpe (2018) has described the discourse of BHT as a "circus-like atmosphere encouraged by Canavero". Wolpe (2017) has noted several rhetorical devices in the Canavero and Ren oeuvre that avoid connecting claims to arguments. One concerns Canavero's call to end animal research because they have been overused and have provided little evidence that BHT will allow full neurological recovery in humans or project how the brain will integrate with the body over time, or even *if* it will integrate. Canavero himself has stated that he does not want to kill any more animals and that they learned everything they needed to know from Robert White's experiments decades earlier. Wolpe, however, contends that the call to abandon animal research was necessary because it did not result in significant medical discoveries that would enable BHT in humans to succeed. In response to this quote by Ren and Canavero (2017: 202) — "It should be abundantly clear that all criticisms raised are groundless as to the feasibility of the project" — Wolpe (2017: 207) further notes:

> It is precisely this continuous dance around the science and medicine that so concerns those that look seriously at Ren and Canavero's claims…. Aside from the patent falsehood of that statement, testing surgical techniques on animals is not intended to lead to a "medical breakthrough," but to demonstrate technical feasibility, expose unexpected complications, and determine the level of success of the desired outcome.

Animal research connected to BHT has contributed little to confirm survival rates, prove spinal cord reconnection and functionality among transplants, project rejection events and control in humans, establish brain-body integration, or demonstrate the potential of mental health distress. A careful read of the animal experiments serves to confirm Wolpe's apprehension of the viability of the surgery.

Reliance on media appearances by the proponents of BHT may be intended to compensate for the lack of acceptance of their scholarly work. Wolpe (2018) has noticed the dearth of published research on BHT techniques and outcomes, and Scott Gelfand (2017) has stated that Ren and Canavero prefer journalists and the popular media to peers when engaging in ethical debates.

In contrast to the numerous media interviews, a search on Google Scholar for medical papers demonstrating or arguing for the feasibility of head transplantation revealed 25 academic papers, editorials, and other books and articles authored or co-authored by Canavero or Ren (See Table 3.1). This is not necessarily an inadequate number of publications for validating proof of concept for a new surgery, and there are likely other publications that the search did not identify; however, an analysis of this bibliography reveals interesting patterns. First, of the 26 titles, 13 appear in *Surgical Neurology International*, whose editorial board includes Sergio Canavero.[5] Three of these papers are short editorials. Another 10 were published in other peer-reviewed journals, and of those, four were published in *CNS Neuroscience & Therapeutics*, three in *Surgery*, two in the *American Journal of Bioethics–Neuroscience* (one is a reply essay), and the final piece is in the *International Journal of Surgery*. One is in Korean and not accessible to English-only speakers, and the two remaining titles were published outside the established peer review system (Amazon and Nova Science Publishing, which has been criticized for its review process).

The academy has mixed opinions on cluster publishing of this nature. While some believe loading the bulk of one's research into one journal is not problematic, others disagree. The opposing view contends that placing the near totality of research in one location has a negative effect on the scholar's reputation and the legitimacy of the research. A suspicion bias, rightly or wrongly, emerges that suggests that the work is of little interest or merit to a wider audience of peers. The rigor of the work is assumed unchallenged and that the topic is only of interest to a limited audience.

In addition, the research conducted on body-head transplantation is quite sparse and is dominated by a single group (Lambda et al. 2016). Unlike face and hand transplant research where several teams in Europe and North America were working largely independently to prepare to conduct those surgeries, there really is only one team conducting head transplant research. The main group in China where Canavero and Ren were based collaborated with a group in South Korea, but the groups were not independent of each other because Canavero's and Ren's names appear as co-authors on the Korean group's papers. The absence of other research teams working on BHT is an indication that "competitors" do not believe it will work. The research that Ren and Canavero have produced lays out the steps of their plan fairly clearly so that another team could take their ideas and conduct their own experiments and test for improvements. That research cannot be found.

---

[5] *Surgical Neurology International* is an open-access, on-line-only journal co-founded by longtime editor James Ausman. The Journal, by online accounts, has an acceptance rate of 50 percent and has been subject to public and academic criticism. One point of contention is that editors allow the journal to serve as an outlet for right-wing political editorials related to neither neurology nor medicine. See for example Ausman and Faria's (2019) essay on gun control, Venezuelan (a *nom de plume* for an anonymous contributor) and Ausman's (2019) editorial on Venezuelan politics, and Faria's review of a book that uses historical Judeo-Christian writings to affirm "the right not only of the homeowner but also any other good citizen, to kill a burglar" (2020: 241).

Table 3.1: General Articles On Body-Head Transplantation Authored By Its Leading Proponents, Derived By A Search On Google Scholar.

1. HEAVEN: The Head Anastomosis Venture Project Outline for the First Human Head Transplantation with Spinal Linkage (GEMINI)
   S Canavero
   Surgical Neurology International, 2013

2. *Head Transplantation and the Quest for Immortality*
   S Canavero
   Amazon CreateSpace IP, 2014

3. Commentary
   S Canavero
   Surgical Neurology International, 2015

4. The "Gemini" Spinal Cord Fusion Protocol: Reloaded
   S Canavero
   Surgical Neurology International, 2015b

5. Human Head Transplantation. Where Do We Stand and a Call to Arms
   XP Ren, S Canavero
   Surgical Neurology International, 2016

6. Central Pain Following Cord Severance for Cephalosomatic Anastomosis
   S Canavero, V Bonicalzi
   CNS Neuroscience & Therapeutics, 2016

7. HEAVEN: the Frankenstein Effect
   S Canavero, XP Ren, CY Kim
   Surgical Neurology International, 2016

8. Report on Long-term Survival after Cervical Reconstruction: The Feasibility of Successful Head Transplantation
   CY Kim, IK Hwang, Canavero S 2016b - dbpia.co.kr. (in Korean)
   https://www.dbpia.co.kr/Journal/articleDetail?nodeId=NODE06617507

9. HEAVEN in the Making: Between the Rock (the Academe) and a Hard Case (A Head Transplant)
   XP Ren, S Canavero
   American Journal of Bioethics - Neuroscience, 2017a

10. First Cephalosomatic Anastomosis in a Human Model
    XP Ren, M. Li, et al.
    Surgical Neurology International, 2017

11. Brain Protection during Cephalosomatic Anastomosis
    XP Ren, EV Orlova, EI Maevsky, V Bonicalzi, S Canavero
    Surgery, 2016

12. Neurologic Foundations of Spinal Cord Fusion (GEMINI)
    S Canavero, XP Ren, CY Kim, E Rosati
    Surgery, 2016

13. Houston, GEMINI has Landed: Spinal Cord Fusion Achieved
    S Canavero, X Ren
    Surgical Neurology International, 2016

Table 3.1 contd. ...

Another method proponents use to control the narrative surrounding BHT, is by inserting BHT into the history of medicine. Head transplantation is often rhetorically referenced as a pioneering historic breakthrough whose developers, like many innovators in medical history, are first misunderstood. Ren and Canavero (2016a) liken BHT to the invention of antiseptic handwashing and balloon angioplasty and to the discoveries of germs, viruses, cancer, helicobacter pylori and ulcers, prions, and the principles of heredity, among others. In this article, they ask, "Why so much acrimony for a lifesaving procedure?" Seating an untried and evidence-shy surgery among the pantheon of some of the most important medical discoveries in all of history is certainly premature, but it seems to be a gesture intended to raise the status of BHT without actually proving its merits.

Why the Narrative Surrounding BHT is an Ethical Matter

Reviewing the narrative surrounding BHT sounds more like the grist of rhetoricians than a subject that would interest bioethicists. Yet, there is ethical substance buried within the context of the discourse. Many of the statements delivered by BHT proponents betray the ethical standards they are presumed to cherish, and hidden within the rhetoric are clues to the shortcomings of the BHT blueprint.

The style in which the ethical argument is presented is wrought with unrelated comparisons that distract from addressing the ethical questions actually posed. For example, in response to charges that BHT violates normative sensibilities of suitability and appropriateness and is therefore repulsive and disgusting to most individuals, a feeling often nicknamed the "yuck factor", Ren and Canavero (2017: 2013) wrote the following:

> As for the yuck factor, we would suggest that humankind never had qualms killing (including nuking), maiming, torturing, starving, or oppressing other humans (Stalin, Hitler ...) and did so for millions of years. We would say that it is humankind that is "both rotten scientifically and lousy ethically."

If one wanted to play word games, one might say that Ren and Canavero are proposing to do no worse than the greatest villains in human history. An interesting, though perhaps awkwardly constructed, syllogism thus follows:

> BHT surgeons are human.
> Humans maim and torture.
> Humans have no qualms about maiming and torturing.
> Therefore, BHT maims and tortures and is acceptable.

Invoking the names of two of recent history's most heinous killers is a simple though telling rhetorical device. On the surface it seems to be saying that BHT is torturous, but why should people worry about it when our entire history is wrought with violence and the inhumane and unthinkable horrors people impose on each other? The main point, however, is that this argumentation is a misdirection. Yes, BHT may be bad, but throughout history people have done much worse. BHT proponents have not addressed authentic ethical criticisms that people may find BHT difficult to tolerate because social values have not prepared them for such a radical medical intervention. Having a donated organ is acceptable throughout most of the world's cultures and recipients, but the idea of an entirely new body that includes genitals, hands, and all new skin may be repulsive to many, if not most, people. Lacking the will or wherewithal to address that question directly is a further indicator of frustration that their ideas have faced such vitriolic resistance in the academic "universe". Of more worry, however, is the trivializing and minimalizing of real human concerns about the possibility of living within another's body. The revulsion may be, and likely is, genuine, and it would be helpful if BHT proponents researched this ethical issue. Such research was conducted on face transplantation acceptability before that surgery became widely conducted (see Barker et al. 2006).

Another point of ethical concern about the BHT discourse is that it conceals the failure of research to prove the viability of the experiment. There is no direct statement that summarizes the animal experiments. Readers are left to do that on their own. In writings that address the robustness of the surgery, proponents do not address that the animals lived for only a short duration and that the neurological functioning of the poor creatures was never fully restored. Nonetheless, proponents have declared numerous times, and with some vigor, that head transplantation is ready for a human trial. Many other facets of the surgery such as anti-rejection protocols, pain control, body-brain incorporation, and psychological stability have also been ignored in the animal research. Animal research has yet to show an outcome that hunans would desire.

Concealing unintended consequences in the rhetoric of BHT violates a standard of bioethics that demands transparency and exhaustiveness in covering all areas of moral and medical threat and concern. The discourse of BHT tries to convince the "audience" that BHT is safe and efficacious when in fact there is little reason to believe it can work. Because proponents cannot show evidence that a patient's health will benefit from a body-head transplant, they can offer little more than hope, which is not in patients' best interests (Gelfand 2017), hope, rather than medical facts, is the centerpiece of their rhetoric. The absence of a coherent and cogent bioethical argument that conforms to the norms of how established science settles its ethical conflicts is in itself a significant error in the ethical profile of head transplantation.

Nationalism and Ethical Compromise

The notion that science is value-free is an obsolete proposition that fails to recognize how science is entwined in the cultural characteristics and social dynamics in which it is conducted and its products are used. Doing science is itself a value because emotional and cognitive decisions, again based on culture, to select science as a means of generating knowledge implies that a person has rejected alternative forms of knowing and generating "truths." Thus, we say that science is value-laden, and any analysis of science, including medicine, must include the web of socio-cultural influences on its direction, character, and utility. No knowledge system operates outside those influences.

A number of examples demonstrate these influences. Insistence of the hegemony of allopathic medicine has caused cultural conflicts between western and non-western cultures and has harmed people's lives. Resistance to psychiatric drugs in South Asian cultures such as Nepal has occurred because the pills did not fit that culture's interpretations of mental illness and were contrary to established treatments provided by the local healer. People did not understand the "magic" of the pills but they did understand the culturally consistent practices of the tribal healers' treatments of herbs, diet changes, and religious incantations (Tausig et al. 2000). Government-funded programs that train local healers to incorporate psychoactive medications into their culturally recognized modes of treatment have proven successful in encouraging people to accept the legitimacy of western treatments and use these medicines to allay their symptoms.

While medicine strives diligently to maintain objectivity and reduce intrusions of physicians' and researchers' emotions, economic motivations, and political opinions into their work, the practice of medicine is wrought with values and subjective interference. Doctors are not immune to the pressures to link their success to their cultural, social, and political foundations. On some occasions, this linkage is unconscious, in others it can be quite deliberate, as the case of Christiaan Barnard's two-headed dog referred to earlier can attest. The patriarchal history of heart disease research kept us from knowing that a heart attack may present differently in women than men, for example. This lag in knowledge was probably an unconscious assumption rooted in precepts of patriarchy rather than intentionally ignoring women's cardiac health. Given the dominant paradigm of male-centeredness that ruled the day, male doctors had no reason to believe that female hearts were different. The question simply did not occur to them.

There are many more examples. To name a few—corporate interests dictating what drugs are developed and how they are priced; insurance companies and for-profit hospitals intruding on the practice of physicians in ways that influence clinical decisions and doctor-patient relationships; doctors whose own biases and prejudices influence how they treat patients of social and ethnic backgrounds different than their own.

If science is social, it is also political. Competition for "bragging rights" and national esteem often hinges on scientific success. Science during the Cold War often garnered as much attention as Olympic athletic competition. Whoever produced the most innovations, just like winning the most gold medals, could jockey for a higher international position and reap new global economic and political partnerships. The winners could also boast that their social and economic systems, and values, were superior to that of the losers. Not lost in this competition was adversaries' race to develop more dangerous weapons and greater means to destroy the other. Science has become an expression of nationalism. Nations bank heavily on developing their scientific prowess and production.

The list of ways in which things social and things medical collide is an old and long one. Inevitably it has reached our story as well.

The Move to China

In 2015, the primary team researching BHT "happily" relocated to China (Ren and Canavero 2016: 240). Sergio Canavero told journalist Bec Crew of Science Alert that the decision to situate the research and to stage the first head transplant there was driven by two factors. First, Canavero had partnered with Xiaoping Ren who was a professor of surgery at Harbin Medical University. Ren had been working on animal head transplant experiments and publishing his results since at least 2013. By 2015, Ren had already performed experiments on over 1,000 rodents. Crew reported that Canavero lauded Ren as the "only person in the world" who could direct the BHT research and lead the surgery when the time came to perform it. Second, Canavero further told Crew that China was ideal for conducting the first BHT because of the country's organizational and operational facilities and capacities. Harbin University awarded Canavero an honorary professorship.

There may been more to the move than resources and a sympathetic ear, however.

Nationalism and medicine are no strangers to one another. Dating back to the turn of the last century, many physicians and scholars have issued warnings of the risks of merging nationalism and medicine. One of the best known treatises on this topic is Henry E. Sigerist's 1947 address given to the conference of the Medico-Chirurgical Society of the District of Columbia, the oldest African-American medical association in the United

States. In his speech, the famed physician and historian of medicine described how nationalism, in opposition to patriotism, in American health care caused discriminatory medical practices and the violation of the neutral delivery of medical care.

More recently, Rebecca Herrero Sâenz (2022) has documented the conjunction of Spanish nationalism and transplant medicine. Owing to an organ shortage in that country, Spain created a campaign to entice individuals to become donors by framing donation as a moral crusade. Herrero Sáenz noted that the movement sought to increase social solidarity through citizen altruism, making it a national duty to care for unknown strangers and fellow Spaniards. Spain emerged from this movement as a leader in transplant medicine, which soon became a part of the country's sense of national unity and pride.

In his analysis of CTA history, Samuel Taylor-Alexander (2013) uncovered nationalistic tendencies in the race to conduct the first face transplants. In more than one country, surgeons became equated as and embodied with national achievement, which led them to view their roles in a national and international context. Medical achievements featured on the global stage translated into national pride for the "powers that be" as well as everyday people who were looking for reasons to feel good about themselves and their nation. The merger of nationalism and medical research, however, carries risks. According to Taylor-Alexander, nationalism's intrusion into face transplantation resulted in "ethical slippage" in which shortcuts were taken and patients who perhaps were not ideal candidates for surgery were selected, and sometimes with catastrophic outcomes.

The 21st Century thus far has witnessed the ascendancy of the Chinese who have flexed their economic, political, and military muscle throughout the world. In their attempts to rival the Europeans and North Americans for recognition as a global power, we see China establishing diplomatic and trade alliances in countries previously aligned with the West, engaging in peacemaking efforts in areas where previously it had little outward signs of influence, and using its vast military resources to extend its boundaries and influence.

There are two types of global influence: "hard power", which refers to objective and material exertions of supremacy and coercion, and "soft power", the capacity to persuade others to follow one's own directions. Whereas hard power, exampled by military action or economic dominance, is coercive and directly manipulative, soft power leads through attractive and desirable qualities that others want to imitate. Music, art, diplomacy, foods, and religion are qualities that may promote one culture to align itself with that of another, thus leading governments and people in one country to see the other in a positive light and awarding them with cultural desirability and authority it may not have otherwise achieved.

Medicine is the stuff of soft power. For over a century, the West has used its medical knowledge to establish political and economic ties with countries throughout the world. The humanitarian efforts of western medical institutions and practitioners have contributed to the esteem of western culture, as well as its power.

This history has not been lost on China. The country has rapidly expanded its medical capacities; however, the country has particularly earmarked medical research and innovation to position itself as a global leader not only in medicine, but in world leadership and favorability as well. On more than one occasion, China as politicized its research. One notable achievement was the first two cloned macaque twins, which were named Zhong Zhong and Hua Hua. Zhonghua literally means Chinese nation and its people.

Canavero and Ren set up their shop in China after their research and BHT aspirations were rebuffed on ethical grounds in both Europe and the United States. Rather than addressing the actual ethical reasons Americans and Europeans withheld approvals to perform his experiments, Canavero responded that "the Americans did not understand" and that "western bioethicists needed to stop patronizing the world", suggesting that westerners were demeaning his work and acting in a patriarchal and narrowminded manner in exercising their authority to deny BHT research and experiments in their countries (see South China Morning Post 2017). Ren and Canavero (2017) have argued that experiments that live on the vanguard of innovation present extraordinary opportunities. Head transplantation, in their thinking, certainly qualifies as extraordinary, and promoting it presents the possibility of a much bigger prize than protecting moralistic orthodoxy, the path taken by the European and American medical establishments. Moral conservativism, BHT researchers contend, can and should be minimized and even discounted because it stems from ignorance, cultural prejudice, and media bias (Shook and Giordano 2017). Established moral values, in this perspective, place restraints on innovative thinkers and retards or even inhibits progress,

China, on the other hand, welcomed Canavero and the head transplant research program perhaps because it would give that country a political one-up on the rest of the world and a clear demonstration of Chinese soft power. A fully successful head transplant would be a medical "man on the moon" achievement and turn all eyes to China and its medical, and thus cultural, superiority. Proponents of BHT recognized China's desire to use medicine to gain international prominence and stature and jumped at the opportunity to work there. Canavero is quoted as saying, "Chinese president Xi Jinping wants to restore China to greatness" (Wolpe 2018).

If the ethical standards of Western medicine have rejected BHT, how did China's ethical system, with its increasingly westernized outlook and structure, decide to accept it? While we cannot know of any closed-door discussions that may have occurred, we can see how it may fit in the grand scheme of bioethics in that country.

China's system of bioethics is grounded on a different set of values than that of most Western nations. Whereas in the West, the individual is generally the basic unit of identity, law, and ideology, in China, it is the family. If a person misbehaves in China, for instance, it is the family that suffers shame and oftentimes an informal punishment. Industrialization and rising educational achievement, however, are creating changes to this sociocultural feature of this society. Still Chinese society remains collectively oriented, which implies that individuals have fewer rights in medical research than they do in the West. Such a worldview discounts the autonomy of an individual body in deference to a collective ideology intending to maintain a strong national body (Cheng 2023).

It is no surprise then that Chinese bioethics have undergone criticism in the last few decades, especially in transplantation medicine. As China developed its medical tourism industry and domestic health care system, the country eventually reached a point in which it could not provide a sufficient number of organs to meet the demand. To resolve this problem, authorities turned to its large supply of executed prisoners to serve as the primary source of transplanted organs (Suah and Mills 2022). Procuring body parts from prisoners was considered ethical in China until the practice was mired by international criticism. By 2010, policymakers abandoned the practice of harvesting prisoners' organs and created new systems of procurement and transplantation based on western models, and an orderly and voluntary system for organ donation was established and standardized across the country. Using prisoners' organs is believed to have ended in 2015 (Shi et al. 2020).

The revised transplantation system, rooted in values of individual permission for donation and democratically fair disbursement of those organs and conducted under the guise of ethical review boards and peer assessment, has not been established throughout the country, however. Methods to review high-risk treatments and experiments are particularly undeveloped. As of 2019 in Zhejiang Province, for example, only 100 of 195 public hospitals, including universities, have their own Institutional Review Boards, but no private hospitals or universities have been approved by the Forum for Ethical Review Committees in the Asian and Western Pacific Region (Ye et al. 2019).

By all accounts, ethical standardization in all medical areas remains inconsistent in China, and research that would be ethically condemned or restricted in the West continues to flourish. One non-transplantation example that made headlines is the scientist in China who altered human genes in three babies to make them resistant to HIV (Ye et al. 2019). Although the scientist was jailed after domestic and international backlash to his pioneering research, questions remain as to how the geneticist was allowed to conduct the research and plant the altered eggs into a womb. Gene-editing research is conducted all over the world, but no one has planted the genetically altered ova because of ethical restraints. Reports indicate that China did toughen its bioethical standards after the scandal.

From an internet search, Harbin Medical University, where Ren and Canavero have conducted their work, appears to have an IRB. A search on the University's English language website, however, yielded no results for searches on the terms "ethics", "bioethics", "review board", "Canavero", or "head transplantation", along with variants of those terms. Of interest, though, is the first statement under the website's Profile tab, which reads:

> Listen to the Party, follow the party, hold high the great banner of Xi Jinping Thought on Socialism with Chinese characteristics for a new era…. (www.hrbmu.edu.cn/english)

It would seem that delivering the nationalistic vision of the government's leaders is a priority for the university, and achieving the world's first human head transplant would be a medical and political triumph for Harbin and the nation.

Chinese bioethics, compared to key tenets held by the Western world, has a number of tension points, however. As Hongqi Wang and Xin Wang (2017) have stated, establishing a code and clinical system of bioethics in China that promotes the three key tenets of the Belmont Report (respect for persons' rights to autonomy, social justice and fairness, and protecting patients and subjects from harm) faces a number of difficulties. First, there remains in China ongoing cultural and social tensions between individual rights and collective rights. As mentioned earlier, in Chinese culture the needs of the collective whole typically outweigh the needs of the individual. The harsh quarantine during the COVID pandemic is a case in point.

Second, Wang and Wang point towards a contradiction that exists between procedural justice and substantive justice in China. Procedural justice maintains that a fair procedure produces a fair outcome. Upholding the specifics of the process, which must be assumed to be equitable and thereby fair, is thus the central focus in ethical behavior when procedural justice is the primary force. Substantive justice refers to the substance of the laws or policies, which must also be perceived as fair. The essence of the ethical codes must be beneficial and fair to all. Substantive justice implies that policies and procedures can be interrupted in the delivery of ethical equity if following the rules in a formal and narrow interpretation leads to hardship or injustice for a person or group. Substantive justice is not achieved when obeying the procedure interferes with "doing what is right". For a just system of bioethics to occur, both the system and its guiding principles must be rooted in beneficence, social justice, and the autonomy of patients.

Often, however, the "turfs" of substantive and procedural justice collide and are out of sync with one another, often taking the form of a conflict of interest. Such problems occur in most places around the world, but in China, the conflicts can be serious and deadly. Wang and Wang cite an example in which physicians were unable to perform a life-saving treatment because the patient was unable to provide consent and the patient's husband, who has authority over the family, refused to allow the treatment. Soon thereafter, the patient died. Following the dictates of the hierarchy took precedence over the physicians' medical decisions, and the medicos' inability to interrupt the rules caused the woman's death. Following the precepts of Chinese collectivism, and the rules, both formal and informal, that maintain it, is often in conflict with emerging norms of individual autonomy and interests. While the trend in both procedure and substance is to move China towards a model of bioethics that centers on individual self-determination, conflicts continue to arise, creating opportunities for physicians to perform innovative, albeit dangerous, experiments. Rules and practice wisdom are often inconsistent.

A last barrier to Chinese bioethics reform is a lack of a global bioethical consensus. A national-universal tension exists that pits Chinese definitions of ethical behavior against rising international standards created by widely accepted formal declarations, such as the Belmont Report, of behavioral expectations among researchers and subjects' rights. While an argument is to be made for global standards of fairness, impartiality, and autonomy to take precedence over cultural-specific and nationalistic goals, China's immediate needs are less hypothetical and more practical. Solving China's intent to solve problems in delivering health care, including transplants, is more pressing than conforming to theoretical models that have their origins outside the culture, according to Wang and Wang.

These conditions have led, relative to Western standards, to the construction of a weak moral or ethical system of medical ethics in which economic and political status determine privilege and compromise the right to do with one's body as one pleases. A relatively loose set of bioethical principles, again looking through Western lenses, set in the context of rising international prominence would welcome the opportunities that body and head transplant could offer. A successful experiment would bring global esteem to China's medical establishment and be a source of national pride for decades, promoting the country's rise to prominence on the international political stage. Given China's governmental control on its academic publications and public media, as well as reporting by international journalists, the research could be done in secret, and report only successes while concealing failures and practices that may be inconsistent with Western bioethics.[6]

BHT and Conflicts with Cultural Narratives

Part of what defines the boundaries between cultures is their unique narrative that specifies its people's sentiments, values, interpretations, and expectations. We know when we leave our own culture and enter another simply by observing the differences in worldviews and perspectives as the people experience everyday life.

Gulfs between the zeitgeists of Western and Confucian models of thinking constitute one of the worlds' greatest cultural divides. Greek particularism opposes Confucian generalism and thus spawn their respective narratives of individualism and collectivism. The discourse that follows these cultural styles has dramatic implications for determining the ethical appropriateness of head transplantation.

As we have seen, BHT raises important questions as to who the person surviving the surgery would be. In the West, the question is set among biological, legal, and existential uncertainties. Biologically, the "new" person will be an amalgam of the two bodies being asked to unite into a single functional whole, but the identity of that new person is likely to follow that of Western individualism, and that identity will be that of the head donor whose memories, perceptions, and personality are (we assume) the ones that will awaken coming out of the prescribed coma. It is fair to say that those social and psychological characteristics will change or certainly be altered in response to the new body, but we will most certainly call BHT patients by the names of the head donor who (again, we assume) will recognize the world as the people they were before the surgery, albeit with a new body. More importantly, Western

---

[6] There is precedence for concealing research and medical practices in China. For one example, at the Second International Symposium on Composite Tissue Allotransplantation in Louisville, Kentucky, in 2000, surgeons from the U.S., Europe, and other countries were discussing the various issues involved in transplanting hands, a surgery still young in its implementation. One of the problems considered was the delay in getting a suitable donor. Surgeons stated that a year was needed to identify a single hand donor and get permission from a cooperative family. A Chinese surgeon, however, surprised the audience by stating that his team had conducted two transplants already (no other team had done more than one at that point) and that finding a donor took hardly any time at all. He did not understand why it took the others so long. The audience was stunned.

Anecdotal accounts attest to that country's many incongruences between policy and practice. While the ethical rules generally conform to Western standards of individual self-determination, the *de facto* reality is that these rules are often fluid in light of social and political pressures and expectations. Norms and behavior do not always match.

BHT survivors will probably see themselves first as the person who was "in" the brain of the head donor.

In East Asian cosmology, however, the cultural narrative has a contrasting theme, which could have major implications for how the surgery's survivors will understand themselves and be seen by other people. Confucian intellectualism portrays human essence as radically different than Western thought and concludes that the person who survives a body-head transplant would be a totally new person, a third person who is neither body donor nor head recipient (Bian and Fan 2021). Why would personal identity be different in China than in a Western culture?

In Greek philosophy, all of nature, including people, is assumed to be composed of parts, which are also made up of smaller parts. The goal of Western science is to separate all substances to determine their unique characteristics and specify the inherent cause and effect relationship between and among them. The assumption is that each substance has its own unique properties and character.

Eastern metaphysics and Chinese traditional medicine do not separate people from nature. Instead, they hold as axiomatic that all substances have properties in common and cannot be differentiated. People and nature exist in a state of harmony, a constant continuity that unites all things as a unified whole. The constant in Confucian philosophy is the concept of *qi* (pronounced chee). *Qi* refers to the universal life energy that is inherent in all things. Just as the Earth and its plants and animals possess *qi*, so do we. *Qi* energizes all the physical systems in our bodies as well as our emotions and thoughts. A healthy and happy person possesses the right qualities of *qi* that exist in the right balance (*yin* and *yang*). *Qi* is who we are. Yet, *qi* can be fluid and if not circulating properly in the body, it can force our essential life energy out of balance. When this happens, distress, both psychological and physiological, ensues. Traditional Chinese medical treatments such as acupuncture are believed to returns *qi* to its proper state, which in turn, restores health.

In Confucian wisdom, people possess their own *qi*, their unique personal life energy that determines who they are. If a head and body are separated from one another, the disruption to the person's *qi* is irreparable. The system of constantly flowing *qi* would be destroyed during a BHT, upsetting the uniqueness of the person and permanently disturbing their life essence. If human life is *qi* activity, then removing the head, for both the head recipient and the brain-dead body donor, halts the flow of *qi* and brings an end to that unique person.

An entire philosophical system is devoted to *qi*, but now is not the time to review it. What is necessary to understand is that *qi* is the defining quality of each person within Chinese metaphysics and centers that country's dominant cultural narrative. In Chinese thought, the body and mind are not separate phenomena or made up on constituent parts, as in the dualism of Western philosophy and medicine. Rather, the body and mind constitute a metaphysical and material whole, a single entity composed of and energized by *qi*.

Furthermore, the notion of harmony extends past the boundaries of individual bodies. The collective unity of the body reaches into social relationships and social organization. Social life follows a similar path of unity. Just as the body strives for harmony and balance within itself and with nature, the highest social virtues are those that promote harmony in social interactions. There is a natural order to rituals, and relationships mandated by heaven unites ancestors and descendants and identifies each person's place in society and their identities. A person living in a Confucian zeitgeist knows who they are by their position within their family's genealogy. Metaphorically, the hierarchy and identity of ancestors and descendants is the DNA of any individual living within the Confucian ethos, making personal identity inseparable from family relationships (Bian and Fan 2021). Separating the body dislocates identity.

Because of *qi*, Bian and Fan argue, BHT is inherently unethical in Confucian metaphysics. To put it succinctly, they state that BHT "would rip apart relationalist features, destroy family relations, and mess up personal identity" (2021: 225). The BHT survivor would be an amalgam of the *qi* of two persons and, therefore, constitute a third person, one who is devoid of identity, essence, and family location.

Jianbui Li and Yaming Li (2022) drew a similar conclusion. Because Confucianism understands selfhood and the body as a singular entity, they cannot be separated—selfhood is embedded within the physical form. Li and Li contend that in a culture rooted in Confucian principles body and head transplantation harms both the person of the body and the person of the head. Great and irreversible damage is done to their existence, changing the selfhood and identity of both. By cultural definition in the Confucian context, a BHT patient is neither the person of the original head or body, but a third person who emerges from destroying the identities and personalities of the individual personages of the head and body.

> Confucian personal virtues greatly depend on selfhood or personal identity, and these depend on a person's body cultivation and mind rectification.... Thus, head transplant would destroy two people's identities and personalities to build a new person (Li and Li 2022: 238).

The person resulting from a BHT is a constructed one, built not of nature or ancestral inheritance but of artificial human action.

As a function of globalization, the world is indeed drifting towards cultural homogeneity. Nonetheless, we are far removed from a time in which all peoples agree on a narrative that explains the essential qualities of life and the nature and precepts of daily living. By definition, in the context of Confucian thought, BHT fails ethically. A culture gap, however, is thus created. It is improper to impose Western ethics on China, yet is the Chinese government imposing its evolving ethical positions, which are moving away from traditional Confucian precepts, on its people? The lives and minds of most Chinese remain fundamentally Confucian, and the thought of disturbing the natural forces of two persons' *qi* and then joining them together would be considered unthinkable and abhorrent.

The Economy of BHT

A pragmatic critique of body-head transplantation is most straightforward when addressing the economics of conducting this experimental surgery. The ethical question, of course, is what price we place on a human life. BHT is indeed intended to be life-saving, but it would occur at an enormous cost. As we discussed earlier, the budget to execute a BHT surgery would reach deep into the tens of millions of dollars. That large team of physicians, nurses, technicians, and support staff, the hospital and its administrators, liability insurance, post-operative care, and various therapies such as psychological, occupational, physical, and pharmaceutical, among others, would put the surgery financially out of reach of most everyone on the planet. No health insurance policy, neither private nor public, would cover the costs for patients, and it's doubtful that the legion of professionals and hospitals and organizations would donate their treasure and time, even for the first attempt.

But how much is too much to save a life? That is an age-old question that when monetized really has no clear ethical answer that would satisfy everyone.

So, let's reframe the question and focus not on money but on another currency: organs and body tissues.

The United Network for Organ Sharing (UNOS) was established by Congress in 1984 to systematize and administer organ procurement and transplantation in the US. As part of its mission UNOS maintains the statistics of need and the supply of donors over a given period of time. As of May 2023, over 100,000 people in the United States were waiting for a life-saving organ transplant, and though the number of donors has increased by 39 percent over the last five years, the need continues to overwhelm the supply of usable organs. Table 3.2 shows UNOS data on organ need.

This list does not tell the entire story. For one thing it only accounts for patients in the United States. If we were to add patients throughout the world in need of an organ transplant, the number would most certainly extend to well beyond a million. In addition, UNOS only reports those patients in need of life-saving organs. Not included are life-enhancing tissues such as bones, tendons, ligaments, skin, heart valves, blood vessels,

Table 3.2: Organ Waiting List, July 31, 2023.

Kidney	88, 676
Pancreas	834
Kidney & Pancreas	1,955
Liver	10,136
Intestine	210
Heart	3,355
Lung	971
Heart & Lung	38
TOTAL	103,527

and corneas, which are the most commonly transplanted tissues in the U.S. according to the CDC (2022).

To be direct, BHT is an inefficient use of scarce organ and tissue stocks and medical labor. Whereas one body might save one life per BHT, the same donor body could save and enhance eight to 15 lives through multiple organ and tissue donations. Furthermore, the time expenditure of physicians, nurses, and technicians dedicated to BHT would be substantial and would distract those professionals from those eight to 15 other patients whose lives could be saved and improved by their expertise.

Another aspect of the insult to medical efficiency that BHT levies is spinal cord repair. If dissected spinal cords could be attached, it would be prudent in economic and humanistic terms to treat the thousands of patients around the world with spinal cord injuries rather than using that technology and the few who possess it to conduct body-head transplants. According to the World Health Organization (2013), 250,000 to 500,000 new cases of spinal cord injuries occur each year worldwide. A more reasonable and humane goal is to reassert people hampered by paralysis into more productive roles and independent lifestyles that improve their quality of life and benefit society. Relieving paralysis would improve self-reliance, release the burden on the health care system and patients' families, and free social resources to address other needs. If PEG enhanced techniques for repairing spinal cord lesions and dissections are proven effective, medical teams and hospitals should devote their time and resources to treating spinal cord injuries rather than focusing on one particular individual.

BHT, on these grounds, not only fails to address current medical needs, but it also directly and intentionally blocks resources from being directed towards meeting those needs.

In this sense, BHT is a morally bankrupt idea. In saving one life, others will be lost, and they are lost with intent. At some point a person with authority would make the decision that one patient acquires all the organs that are available from one donor *and* deny eight or more other patients access to those same organs. At the most human level, this is an impossible decision to reach, especially on an ongoing basis should BHT prove successful. Who would make the decision as to who lives and who dies? And what criteria would justify that decision? There are no medical principles in a democratic system of fair, unbiased health care that would or could warrant such a decision. BHT is a selfish enterprise that would benefit a few people over many others. BHT should be denied on the singular grounds as an inefficient use of organic resources.

Animal Research

As unsavory as it may be, research on animals prior to approving a surgery or a drug for human application is necessary to prevent harm to humans. Medicine's long history of testing new procedures and pharmaceuticals on animals has saved countless human lives. Unfortunately, saving those lives is forever attached to a sordid and often vile process of torture and abuse of creatures both great and small. Over the last several decades, animal rights movements have had a profound impact on the perception of animals and protecting their quality of life. Animals have more protections now than ever in history, and societies are advancing those rights at a quickening pace. Indeed, within the context of advancing empathy towards animals, medical and legal institutions have changed their valuative stances on animal rights. These changes are visible in a number of ways. In the United States, fewer people are interested in hunting, the pet population has increased and the pet care industry has never been larger, and laws have been enacted to protect animals from abuse and endangerment.

Changing attitudes towards animals are also present in medical research. A report by the prestigious Hastings Center (Conlee and Rowan 2012) describes the scarcity of rhesus macaques available for medical research. This shortage (there are about half as many of these monkeys available for research as there were in 1970) is a direct consequence of protecting them from harm.

Animals used in research have their own ethical protections as research subjects. All studies involving animals in the U.S. must be cleared by their IRB equivalents—Institutional Animal Care and Use Committees (IACUCs). IACUCs are responsible for ensuring that experimental animals are treated humanely, housed in clean and comfortable facilities, are well fed, and that researchers minimize pain and distress. These standards are applied to all animals used in research regardless of their species.

Xiaoping Ren and his colleagues have conducted head attachments (placing a second head onto animals) and transplants on hundreds if not thousands of mice, rats, dogs, and monkeys. His team has even conducted cross-species grafts, transplanting a mouse head onto a rat's body, a fact easier to locate in public media via the internet than in scholarly journals (Gelfand 2017; Baldwin 2017). Ren, Canavero, and their associates have claimed their experiments are not only successful but constitute proof of concept despite the reality that their animals' post-experimental lives were measured in hours and days. Some have lived longer, but they are only a few of the countless creatures exposed to such experiments. In one study, for example, Ren and colleagues (2015) declared success for a study on mice in which 12 lived longer than a day and an average of 36 hours and displayed pain and corneal reflexes. The experiment did prove that blood flow from the body to the new head and that some neurological functioning are possible. These accomplishments, however, do not demonstrate that a human's life will be improved or saved after a body-head transplant (Gelfand 2017).

But that is not necessarily the point here. What is of immediate interest is the treatment of the animal subjects. The studies do not report on any pain the animals may have exhibited, nor is there a report on how the pain, if any, was treated. Were the animals suffering severe central pain as predicted? Canavero has openly disagreed with his partner's continuation of animal research, saying that Robert White's early studies rendered all the knowledge possible from transplanting animal heads. As was stated in an earlier chapter, witnesses testified to the immense pain that White's animals were forced to endure. Standards have changed since White's time, but Ren's studies have been conducted in jurisdictions where ethical standards are relatively lax and animal rights are less protected.

Relocating their research in China, as Ren and Canavero have done, allowed them to take research liberties that would likely be denied to them in the West. China has no general laws shielding animals from cruelty, but legal protections for laboratory animals do exist. These laws, however, are lax and subject to interpretation (Cao 2018). China's laboratories remain a major consumer of animal research subjects. By some estimates, each year over 12 million animals are used in medical and cosmetic testing in China, 40,000 of them are primates, and the country remains a major exporter of primates to labs in other countries (Rogers 2019). The poor regulation and lack of endorsement of animal rights allows BHT researchers to conduct excessive and needless experimentation on animals in violation of ethical standards adopted throughout much of the world.

Whole Body Integration and Risks to Identity

As we discussed earlier, BHT proponents approach the psychological aspect of body-head transplantation from a cerebrocentristic perspective, which contends that the mind is a function of the brain and develops independently of the body. In this view, the head and brain can freely move from body to body without disturbing the psychological well-being of the individual. This perspective, however, ignores and dismisses volumes of research in biology, psychology, and sociology that has established that:

- Non-brain body functions (the gut-biome, for instance) influence the brain and mental health.
- Social forces (e.g., poverty or wealth) impact individuals' sense of self and identity
- Social forces impact the body (mental vs. manual labor) and perceptions of the body (racism and stigma).
- The nonmaterial aspects of the mind (distress) affect the body (stress reactions; substance abuse), which can distort self-image and self-esteem.

Disturbing the established brain and body connections and their intersections with identity and selfhood is more than a medical problem; it is also an ethical barrier to performing BHTs. Is it morally right to perform a surgery that in addition to creating massive health risks also endangers the stability and continuation of the head donor's personality and identity?

Anto Čartolovni and Antonio Spagnolo, who were among the first to critique the then new idea of head transplantation, were quick to point out the psychological difficulties BHT patients would be likely to experience as they attempted to incorporate a new body in their "already existing body schema and body image" (2015: 103). Čartolovni and Spagnolo were convinced that memories of the old body's role in personal identity and the absence of that body would lead to such extreme psychological conflict that insanity would be inevitable and eventually lead to death. Madness may seem an extreme prediction on the surface, yet it is not so easily dismissed.

A full body replacement is not equivalent medically or psychologically to a single or double organ transplant or any composite tissue transplant such as a face or hand. The existing body can absorb organs and tissues without significant changes in overall body function or undermining psychological stability. Individuals receiving an organ, and most certainly a face, experience a wide range of emotional changes after the transplant. Many feel a strong sense of gratitude, hopefulness, and joy, though others may develop symptoms of depression and anxiety. One study of kidney recipients found that about one in five patients who received their kidney from a dead donor demonstrated clinically significant depression and anxiety, which were worsened in cases in which immunosuppression problems occurred (Zimmerman et al. 2016).

Emotional responses are not the only "mental" action in which transplant patients engage. All transplant patients engage in existential self-reflection when facing death and receiving an organ that originated in another person (Svenaeus 2012). As Fredrik Svenaeus clarifies, the qualities of the organs determine their existential significance, and patients' existential confrontations with their transplants varies by the meanings and interpretations of those qualities. Svenaeus contends that the ability to touch and see the new organ calls attention to and amplifies the organ being "not me" and impacts recipients' ability to embrace or, using Svenaeus' word, harbor a part of the donor's identity.

Svenaeus also notes a degree of alienation is common to all transplants. With organs that are less connected to identity, such as a kidney or liver, the process of alienation is not particularly impactful. These transplanted organs produce a sense of bodily alienation in that a part of recipients' bodies that does not belong to them, but they do not necessarily cause a disruption to their sense of self.

In other instances where the transplanted organ is entwined with the identity of the donor, the alienation process can be more dramatic:

> In cases in which the organ in question is taken to harbor the identity of another person to a larger extent, because of its symbolic qualities (the heart) or its expressive qualities (the hand and the face), the alienation process also involves the otherness of another person making itself (at least imaginatively) known (Svenaeus 2012: 155).

The more visible the transplanted organ or tissue and the greater social symbolic meaning attached to the organ, the more intrusive the identity of the other person will be for the recipient's own adjustment to the transplant.

The psychological impact of these surgeries is minor compared to what a BHT patient would experience. BHT implies transplanting *every* organ below the face and brain. The transplants are simultaneously visible and hidden, symbolically relatively insignificant and highly intimate. The head recipient's identity would likely not focus too much on an organ that holds little social import, but the combined effect of *every* such organ could be enormous especially when combined with those that are carry social and expressive value such as skin, hands, heart, and genitals. If the transplant of a single organ that lacks symbolic and expressive meaning carries existential meaning and reflection, taking in an entire body could amount to a potentially catastrophic existential crisis.

The ability of the brain and body to integrate with each other does not invoolve just medically connecting the dots. BHT implies a complete reorganization of the brain's relationship to its body, not one or two organs as in a usual transplant. Not only must the brain and body connect, the scenario in which an organ failure anywhere in the new body must be avoided lest the entire BHT fail.

Equally important is that successful body and head integration would have implications for psychological well-being as well. As with face and hand transplants, a total pre-operative psychological assessment would certainly be required before declaring a patient suitable for a BHT. What that assessment would tell us, however, is only a theoretical, not evidentiary, probability of risk to psychological continuation and health. A full battery of psychological testing could identify a patient unlikely to handle the psychological impact of a new body, but there are serious doubts that testing would determine if a person could adjust to a new body without psychological fallout. In other words, a psychological evaluation (probably) would tell who could not handle it, but it would not say who could. At best, testing would give us a probability of adjustment, though there is no data on which to normalize that predictive statistic.

The best predictor of future behavior, of course, is past behavior. An evaluator would most certainly be interested in identifying a patient's history of mood disorders such as depression and anxiety, personality disorders, and serious mental illness like schizophrenia and bipolar disorders. A testing psychologist would also look for indicators of coping skills and how a potential patient has managed acute and chronic stress in the past.

Pragmatically speaking, this stratagem is where a psych assessment would fall short. The brain's response to communicating with a new body and the mind subjectively accepting it are not acute stressors, they are traumas, and severe ones. BHT would cause such an extraordinary disruption in ordinary body functions and shift in body image that patients are more likely to experience post-traumatic stress than they are a continuation of their pre-operative mental health, self-esteem, and identity.

Short of Čartolovni and Spagnolo's prediction of insanity and death, the more likely outcomes are one or more of the following:

- depression and anxiety derived from the persistent pain, paralysis, and slow rehabilitation.
- post-traumatic stress disorder caused by the severe shock to the body from the failure of the brain to coordinate its neurological functions with the new body.
- body dysmorphic disorder, a serious mental illness in which a person exhibits high anxiety about their appearance.
- bodily dissociation, also a serious mental illness in which individuals dissociate from their own bodies.

A condition that is most worrisome ethically because it is associated with trauma and chronic pain is bodily dissociation, also known as self-depersonalization. Bodily dissociation is characterized by difficulty identifying emotions, feeling frozen or numb, and perhaps most importantly in the case of BHT, feeling separated from one's own body (Price and Thompson 2007). At the first sign of any of these psychological conditions, mental health professionals would need to maintain surveillance for suicidal ideation and behavior. If depression and dissociation became severe, patients may feel they need to kill their bodies to free themselves from the pain, physical and emotional, that envelopes them.

A medically successful body-head transplant cannot assume psychological success. The subjective self can reject the new body such as easily as the objective immune system. Dissatisfaction with the appearance or the mere emotional incongruence between self-identity and body image with that of the new body has the potential to threaten the success of the surgery. It is not expected to be easy to help a patient adjust to a body that once belonged to someone else, a body that can be touched, judged visually, and even tasted. Having someone else's genitals is not comparable to having another's kidney or even heart. Those latter organs cannot be touched and do not have the same histories as genitalia. An "I don't know where that's been" interpretation of external sex organs may cause a strong sense of disgust and revulsion that the patient might find hard to overcome. Emotional rejection and disassociation could easily follow.

The possibility of a dramatic emotional response requires preparation of a herculean therapeutic treatment plan and highly skilled mental health professionals to prevent a patient from emotionally rejecting the body. If the risks of psychological trauma are so high, a detailed plan of intervention must be presented along with the surgical protocol. The psychological interventions proposed by BHT advocates, however, is, to put it bluntly, hardly heroic. They have proposed a psychological treatment plan that rests on two strategies, neither of which is proven effective nor routinely used by psychiatrists, psychologists, or clinical social workers.

Ren and Canavero described their strategy for responding to any psychological disturbance in body image by stating they planned to employ "immersive virtual reality" (IVR) before and after the surgery "so that the subject can grow used to the new body" and with hypnosis "to enhance the adaptation process" 2017: 202). Neither of these techniques is established courses of practice in psychiatry and psychology, and neither are standard care for the cognitive dissonance and emotional reactions that may arise after the surgery. IVR holds promise as a clinical tool, but it is, at best, experimental, and cautions have been raised to tread lightly when generalizing laboratory results to real-world interactions (Vasser and Aru 2020). At this point in its development, there is no evidence to warrant entrusting IVR to something as important as the psychological well-being of body-head transplant patients. The risks are too extreme.

The same can be said of hypnosis. There is nothing magical or mysterious about hypnosis; it is a deep relaxation or mindfulness in which patients can separate themselves from their usual patterns of thinking and sensory reactions. Individuals under hypnosis enter into a deep state of centeredness and attentiveness. Just as athletes and musicians get into "the zone", a profound state of focus that allows them to concentrate fully on their performance, hypnosis patients enter into a "zone" in which they exert control over their thoughts and emotions. Hypnotists cannot force patients to do things against their will such as cluck like a chicken as in old and farcical vaudeville routines. Quite the opposite: hypnosis theoretically leaves patients in greater control of themselves.

Hypnosis has practical applications for recovering from surgery. It can help lessen mental distress, pain, medication consumption, and surgical procedure time (Holler et al. 2021). On the other hand, hypnosis is not necessarily a reliable technique in clinical settings because not all individuals are responsive and amenable to "going under". Hypnosis cannot be relied upon to improve the well-being of all patients, and it is unclear who and under what conditions hypnosis may be effective (Holler et al. 2021). Furthermore, it must be emphasized that hypnosis is an adjunct therapy intended to enhance standard post-surgical medical and psychological care. While for some mental health professionals, hypnosis might be considered a necessary technique for post-operative recovery, it is not sufficient for handling the pain and adjustments that often follow surgery. It is certainly insufficient for assimilating a new body into one's sense of self. For hypnosis to be mentioned as one of two techniques specified by BHT proponents for mental health care is shortsighted, deceptive, and an over-simplification of the risks involved to identity and psychological well-being.

A reliance on virtual reality and hypnosis constitutes little more than a crude response to a complex problem. The promoters of body-head transplant are relying on experimental and adjunct techniques to address the dysmorphic and traumatic disturbances patients are likely to experience. Denying the possibility of severe psychological distress and failing to have in place a well-conceived clinical plan to address all possible outcomes is an ethical shortcoming of the BHT protocol. Furthermore, and more importantly, exposing patients to such extreme psychological danger is a violation of the Nuremburg Code which clearly states that an experiment should avoid all unnecessary physical and mental suffering and injury. The degree of psychological risk inherent in BHT exceeds the humanitarian benefits that the procedure could possibly yield.

Conclusion

Body-head transplantation fails to satisfy the standards set by established authorities that monitor and regulate medical interventions and research. As we have demonstrated here, there is little reason to believe that these authorities are acting with any malice or in a needlessly overly restrictive manner. On its own merits, BHT falls short on many ethical criteria.

That there are unresolved technical problems that undermine the call to conduct the surgery is itself an ethical alarm that should prohibit the surgery from being attempted. We must conclude that BHT is questionable ethically on two fronts: doubts that it will work and the rhetoric surrounding it. It seems morally dubious to promote a medical intervention that has no supportive evidence as to its effectiveness and safety. The whole atmosphere of the BHT narrative does indeed feel like a circus, as many have described it. And while it is often true that efficacious innovations are initially met with resistance before they are adopted, the diffusion of new technology and ideas hinges on their veracity. Causing paradigm shifts is hard business, but in science they are not likely to change without proof. Animal studies have yet to produce a transplant in which full neurological functioning was restored, and there is really no way to study or predict the psychosocial aftermath. Generalizing these lab studies and theories to actual human beings is farfetched, a stretch of logic. The entire enterprise of body-head transplantation is based on inference and not confirmation.

I see no real situation in which BHT could be conceived as ethical. Even if the two donors had a special relationship with each other, were medically matched, and gave consent in advance, the head donor is still taking risks that are dangerous and perhaps not survivable. More evidence is needed to make this a legitimate medical option for patients with terminal diseases.

Solving the medical problems would not necessarily migrate BHT from morally wrong to legitimate. A number of ethical fault lines would persist. It is especially hard to get past the selfishness of hoarding donatable organs and tissues and denying them to other equally desperate patients. If the organs dedicated to one BHT could save eight lives, two would save 16, and three would save 24. If conceding eight lives to save one is seen to be morally and ethically viable, then it "ought not be".

References

Asusman, J.I. and M.A. Faria. 2019. Is gun control really about people control? Surgical Neurology International, 10: 195. doi.org/10.25259/SNI_480_2019

Baldwin, E. 2017. An outlandish surgeon who aims to perform the first head transplant just gave a rat a second head. Business Insider. businessinsider.com/head-transplant-rat-experiment-2017-4 (accessed 14 June 2023)

Barker, J.H., A. Furr, M. Cunningham, F. Grossi, D. Vasilic, B. Storey et al. 2006. Investigation of Risk Acceptance in Facial Transplantation: Plastic and Reconstructive Surgery, 118(3): 663–670. doi.org/10.1097/01.prs.0000233202.98336.8c

Bian, L. and R. Fan. 2022. Who Would the Person Be after a Head Transplant? A Confucian Reflection. The Journal of Medicine and Philosophy: A Forum for

Bioethics and Philosophy of Medicine, 47(2): 210–229. doi.org/10.1093/jmp/jhab024

Canavero, S. 2013. HEAVEN: The head anastomosis venture Project outline for the first human head transplantation with spinal linkage (GEMINI). Surgical Neurology International, 4(2): 335. doi.org/10.4103/2152-7806.113444

Canavero, S. 2014. Head Transplantation and the Quest for Immortality. Amazon. CreateSpace Independent Publishing Platform.

Canavero, S. 2015a. Commentary. Surgical Neurology International, 6: 103.

Canavero, S. 2015b. The "Gemini" spinal cord fusion protocol: Reloaded. Surgical Neurology International, 6(1): 18. doi.org/10.4103/2152-7806.150674

Canavero, S. 2022b. Whole brain transplantation in man: Technically feasible. Surgical Neurology International, 13: 594. doi.org/10.25259/SNI_1130_2022

Canavero, S. 2022b. Whole brain transplantation in man: Technically feasible. Surgical Neurology International, 13: 594. doi.org/10.25259/SNI_1130_2022

Canavero, S. and V. Bonicalzi. 2016. Central pain following cord severance for cephalosomatic anastomosis. CNS Neuroscience & Therapeutics, 22(4): 271–274. doi.org/10.1111/cns.12527

Canavero, S. and X.P. Ren. 2016a. The spark of life: Engaging the cortico-truncoreticulo-propriospinal pathway by electrical stimulation. CNS Neuroscience & Therapeutics, 22(4): 260–261. doi.org/10.1111/cns.12520

Canavero, S. and X.P. Ren. 2016b. Houston, GEMINI has landed: Spinal cord fusion achieved. Surgical Neurology International, 7(25): 626. doi.org/10.4103/2152-7806.190473

Canavero, S., X.P. Ren and C.Y. Kim. 2016. HEAVEN: The Frankenstein effect. Surgical Neurology International, 7(25): 623. https://doi.org/10.4103/2152-7806.190472

Canavero, S. and X.P. Ren. 2019. Advancing the technology for head transplants: From immunology to peripheral nerve fusion. Surgical Neurology International, 10, 240. doi.org/10.25259/SNI_495_2019

Canavero, S. and X.P. Ren. 2020. The Technology of Head Transplantation. Nova Science Publishers.

Canavero, S., X.P. Ren and C.Y. Kim. 2017. Reconstructing the severed spinal cord. Surgical Neurology International, 8(1): 285. https://doi.org/10.4103/sni.sni_406_17

Caplan, A. 26 February 2015. Doctor seeking to perform head transplant is out of his mind. Forbes. forbes.com/sites/arthurcaplan/2015/02/26/doctor-seeking-to-perform-head-transplant-is-out-of-his-mind/?sh=5f86be765ed3 (accessed 3 May 2023)

Caplan, A. 13 December 2017. Promise of world's first head transplant is truly fake news. Chicago Tribune. chicagotribune.com/opinion/commentary/ct-perspec-head-transplant-ethics-1215-story.html (accessed 30 August 2022)

Cao, D. 2018. Ethical questions for research ethics: Animal research in China. Journal of Animal Ethics, 8(2): 138–149. doi.org/10.5406/janimalethics.8.2.0138

Čartolovni, A. and A.G. Spagnolo. 2015. Ethical considerations regarding head transplantation. Surgical Neurology International, 6: 103.

Centers for Disease Control. 2022a. Transplant Safety. Centers for Disease Control. cdc.gov/transplantsafety/overview/key-facts.html (accessed 23 April 2023)

Cheng, Y. 13 April 2018. China will always be bad at bioethics. Foreign Policy. foreignpolicy.com/2018/04/13/china-will-always-be-bad-at-bioethics/ (accessed 8 May 2023)

Cherry, M.J. 2022. What happens if the brain goes elsewhere? Reflections on head transplantation and personal embodiment. The Journal of Medicine and Philosophy: A Forum for Bioethics and Philosophy of Medicine, 47(2): 240–256. doi.org/10.1093/jmp/jhab045.

Conlee, K.M. and A.N. Rowan. 2012. The case for phasing out experiments on primates. Hastings Center Report, 42(s1): S31–S34. doi.org/10.1002/hast.106

Corrigan, O. 2003. Empty ethics: The problem with informed consent. Sociology of Health and Illness, 25(7): 768–792. doi.org/10.1046/j.1467-9566.2003.00369.x

Cuoco, J.A. 2016. Reproductive implications of human head transplantation. Surgical Neurology International, 7(48).

DiSilvestro, R., C. Choe-Smith, T. Houk and S. Ayala-Lopez. 2017. The road to HEAVEN is paved with good intentions: Transplanting heads, manipulating selves, and reassigning genders. AJOB Neuroscience, 8(4): 223–225. doi.org/10.1080/21507740.2017.1392383

Farahany, N. 25 August 2016. Can you legally consent to a head transplant? The Washington Post. washingtonpost.com/news/volokh-conspiracy/wp/2016/08/25/can-you-legally-consent-to-a-head-transplant/?noredirect=on&utm_term=.acf6ffb5f76b (accessed 12 June 12, 2023)

Faria, M. A. 2020. The moral philosophy of self-defense and resistance to tyranny in the Judeo-Christian Tradition – A review of David Kopel's The Morality of Self-defense and Military Action: The Judeo-Christian Tradition (2017). Surgical Neurology International, 11, 241. doi.org/10.25259/SNI_436_2020

Furr, A. 2022. The Sociology of Mental Health and Illness. Sage, Los Angeles.

Gelfand, S. 2017. HEAVEN, equipoise, and what's best for the patient. AJOB Neuroscience, 8(4), 219–221. doi.org/10.1080/21507740.2017.1392376

Herrero Sáenz, R. 2022. The Media Discourses on Organ Donation and Transplantation in Spain (1954–2020) and their Implications for Spanish Nationalism. Ph.D. Dissertation, State University of New York at Albany. spot.lib.auburn.edu/login?url=https://www.proquest.com/dissertations-theses/media-discourses-on-organ-donation/docview/2712752714/se-2

Holler, M., S. Koranyi, B. Straus and J. Rosendahl. 2021. Efficacy of hypnosis in adults undergoing surgical procedures: A meta-analytic update. Clinical Psychology Review, 85, 102001. doi.org/10.1016/j.cpr.2021.102001

Iltis, A. 2022. Heads, bodies, brains, and selves: Personal identity and the ethics of whole-body transplantation. The Journal of Medicine and Philosophy: A Forum for Bioethics and Philosophy of Medicine, 47(2): 257–278. doi. org/10.1093/jmp/jhab049

Kim, C.Y., I.K. Hwang and S. Canavero. 2016b. Report on long-term survival after cervical reconstruction: The feasibility of successful head transplantation. dbpia.co.kr. (in Korean) dbpia.co.kr/Journal/articleDetail?nodeId=NODE06617507 (accessed 29 July2023)

Kuk, J.H. and Y.J. Ryu. 2019. Critical review of the "head transplant" surgery plan. Journal of the Korea Bioethics Association, 20(2): 45–58. 10.37305/JKBA.2019.12.20.2.45

Lamba, N., D. Holsgrove and M.L. Broekman. 2016. The history of head transplantation: A review. Acta Neurochirurgica, 158(12): 2239–2247. doi. org/10.1007/s00701-016-2984-0

Li, J. and Y. Li. 2022. The ethics of head transplant from the Confucian perspective of human virtues. The Journal of Medicine and Philosophy: A Forum for Bioethics and Philosophy of Medicine, 47(2): 230–239. doi.org/10.1093/jmp/ jhab051

Li, P.W., X. Zhao, Y.L. Zhao, B.J. Wang, Y. Song, Z.L. Shen et al. 2017. A cross-circulated bicephalic model of head transplantation. CNS Neuroscience & Therapeutics, 23(6): 535–541. doi.org/10.1111/cns.12700

Liu, Z., S. Ren, K. Fu, Q. Wu, J. Wu, L. Hou et al. 2018. Restoration of motor function after operative reconstruction of the acutely transected spinal cord in the canine model. Surgery, 163(5): 976–983. doi.org/10.1016/j.surg.2017.10.015

Maisel, E.R. 2013. The new definition of a mental disorder: Is it an improvement or another brazen attempt to name a non-existing thing? Psychology Today. psychologytoday.com/us/blog/rethinking-mental-health/201307/the-new-definition-mental-disorder (accessed 23 March 2019)

Mirkes, S.R. 2018. Human head transplants: Why it's time for a serious debate. Ethics and Medicine, 34(3): 163–169.

Osborne, H. 17 November 2017b. First human head transplant successfully performed on corpse, Sergio Canavero announces. Newsweek. newsweek. com/first-human-head-transplant-corpse-sergio-canavero-714649

Petrini, C. 2013. Surgical experimentation and clinical trials: Differences and related ethical problems. Annali Dell'Istituto Superiore Di Sanità, 49(2): 230–233. doi.org/10.4415/ANN_13_02_14

Price, C.J. and E.A. Thompson. 2007. Measuring dimensions of body connection: Body awareness and bodily dissociation. The Journal of Alternative and Complementary Medicine, 13(9): 945–953. doi.org/10.1089/acm.2007.0537

Ren, S., Z. Liu, C.Y. Kim, K. Fu, Q. Wu, L.T. Hou et al. 2019. Reconstruction of the spinal cord of spinal transected dogs with polyethylene glycol. Surgical Neurology International, 10, 50. https://doi.org/10.25259/SNI-73-2019

Ren, X.P. 2016b. The age of head transplants. CNS Neuroscience & Therapeutics, 22(4): 257–259. https://doi.org/10.1111/cns.12526

Ren, X.P. and S. Canavero. 2016. Human head transplantation. Where do we stand and a call to arms. Surgical Neurology International, 7(1): 11. doi. org/10.4103/2152-7806.175074

Ren, X.P. and S. Canavero. 2017a. HEAVEN in the making: Between the rock (the academe) and a hard case (a head transplant). AJOB Neuroscience, 8(4): 200–205. doi.org/10.1080/21507740.2017.1392372

Ren, X.P. and S. Canavero. 2017b. The new age of head transplants: A response to critics. AJOB Neuroscience, 8(4): 239–241. doi.org/10.1080/21507740.2017. 1393028

Ren, X.P. and S. Canavero. 2017b. From hysteria to hope: The rise of head transplantation. International Journal of Surgery, 41: 203–204. https://doi. org/10.1016/j.ijsu.2017.02.003

Ren, X.P., C.Y. Kim and S. Canavero. 2019. Bridging the gap: Spinal cord fusion as a treatment of chronic spinal cord injury. Surgical Neurology International, 10, 51. doi.org/10.25259/SNI-19-2019

Ren, X.P., M. Li, X. Zhao, Z. Liu, S. Ren, Y. Zhang, S. Zhang and S. Canavero. 2017. First cephalosomatic anastomosis in a human model. Surgical Neurology International, 8(1): 276. doi.org/10.4103/sni.sni_415_17

Ren, X.P., E.V. Orlova, E.I. Maevsky, V. Bonicalzi and S. Canavero. 2016. Brain protection during cephalosomatic anastomosis. Surgery, 160(1). 5–10. https://doi.org/genini

Rogers, O. 2019. Animal research in China. Faunalytics. faunalytics.org/animal-research-in-china/ (accessed 28 June 2023)

Shi, B.Y., Z.J. Liu and T. Yu. 2020. Development of the organ donation and transplantation system in China. Chinese Medical Journal, 133(7): 760–765. doi.org/10.1097/CM9.0000000000000779

Shook, J.R. and J. Giordano. 2017. Ethics transplants? Addressing the risks and benefits of guiding international biomedicine. AJOB Neuroscience, 8(4): 230–232. doi.org/10.1080/21507740.2017.1392377

South China Morning Post. 18 November 2017. World's first head transplant to be performed in China soon, says 'radical' US surgeon. South China Morning Post.scmp.com/news/world/europe/article/2120503/worlds-first-head-transplant-be-performed-china-soon-says-radical (accessed 31 July 2022)

Suah, A. and M. Mills. 2022. Ethics and National Health Policy Change: A Case Study of the Transplant System in China. pp. 605–614. *In*: V.A. Lonchyna, P. Kelley and P. Angelos [eds.]. Springer Nature, Cham, Switzerland.

Suskin, Z.D. and J.J. Giordano. 2018. Body–to-head transplant; a "caputal" crime? Examining the corpus of ethical and legal issues. Philosophy, Ethics, and Humanities in Medicine, 13(1): 10. doi.org/10.1186/s13010-018-0063-2

Svenaeus, F. 2012. Organ transplantation and personal identity: How does loss and change of organs affect the self? Journal of Medicine and Philosophy, 37(2): 139–158. doi.org/10.1093/jmp/jhs011

Taylor-Alexander, S. 2013. On face transplantation: Ethical slippage and quiet death in experimental biomedicine. Anthropology Today, 29(1): 13–16. doi.org/10.1111/1467-8322.12004

Tausig, M., S. Subedee, J. Subedee, J. Ross, C.L. Broughton, R. Singh et al. 2000. Mental illness in Jiri, Nepal. Contributions to Nepalese Sociology. The Jiri Issue: 200-215.

Thomas, W.I. and D.S. Thomas. 1928. The Child in America: Behavior Problems and Programs. Knopf, New York.

UNOS. 2023. Actions to strengthen the U.S. organ donation and transplant system. United Network for Organ Sharing. unos.org/data/(Accessed 23 May 2023)

Van Assche, K. and A. Pascalev. 2018a. Full body transplantation, is it allowed? Transplantation, 102(Supplement 7): S223. doi.org/10.1097/01.tp.0000542886.27707.45

Van Assche, K. and A. Pascalev. 2018b. Where are we heading: The legality of human body. Issues in Law and Medicine, 33(3): 3–20.

Vasser, M. and J. Aru. 2020. Guidelines for immersive virtual reality in psychological research. Current Opinion in Psychology, 36: 71–76. doi.org/10.1016/j.copsyc.2020.04.010

Venezuelan, A. and J. Ausman. 2019. The devastating Venezuelan crisis. Surgical Neurology International, 10, 145. doi.org/10.25259/SNI_342_2019

Wang, H. and X. Wang. 2015. Medical ethics education in China. pp. 81–92. *In*: H.A.M.J. ten Have [ed.]. Bioethics Education in a Global Perspective. Advancing Global Bioethics, Vol. 4. Springer Netherlands. doi.org/10.1007/978-94-017-9232-5_7

Wells, S. 2023. Head Transplant Surgeon Claims Human Brain Transplants Are "Technically Feasible." vice.com. vice.com/en/article/g5v7gy/head-transplant-surgeon-claims-human-brain-transplants-are-technically-feasible accessed (accessed 2 May 2023)

Wolpe, P.R. 2017. Ahead of our time: Why head transplantation is ethically unsupportable. AJOB Neuroscience, 8(4): 206–210. doi.org/10.1080/21507740.2017.1392386

Wolpe, P.R. 12 June 2018. A human head transplant would be reckless and ghastly. It's time to talk about it. Vox. vox.com/the-big-idea/2018/4/2/1717d3470/human-head-transplant-canavero-ethics-bioethics (accessed 31 July 2022)

World Health Organization. 2013. Spinal Cord Injury (Accessed 4 October 2015). who.int/mediacentre/factsheets/fs384/en/ (accessed 12 March 2023)

Ye, Z.J., X.Y. Zhang, J. Liang and Y. Tang. 2020. The challenges of medical ethics in China: Are gene-edited babies enough? Science and Engineering Ethics, 26(1): 123–125. doi.org/10.1007/s11948-019-00090-7

Zimmermann, T., S. Pabst, A. Bertram, M. Schiffer and M. De Zwaan. 2016. Differences in emotional responses in living and deceased donor kidney transplant patients. Clinical Kidney Journal, 9(3): 503–509. doi.org/10.1093/ckj/sfw012

4

Reaching Conclusions
Final Thoughts on Embodiment and Body-Head Transplantation

Chapter Summary

Chapter 4 draws conclusions as to the ethical grounding of body-head transplantation. Based on the risks that have not been fully explored in research, the methods in which BHT has been promoted, and the failure to account for the psychosocial possibilities of a person becoming "someone else" by virtue of acquiring 80 percent of their body mass from another person, the final summation of this procedure is that now is not the time to do it. In truth, there may never be a time or reason to do a BHT. The economics of BHT alone are "deal-breakers".

This chapter counters arguments that BHT patients and face transplantees would essentially be no different, an idea proffered by BHT proponents. It is not just that the surgeries hold few similarities, but also the impact on the embodied personhood of the patient is also not comparable. In this chapter, we offer a few last thoughts about the meaning of the body to show that losing and gaining a body are not likely to be passive exercises in post-operative recovery. They are likely to have severe and negative psychological effects.

* * *

In a perfect world, the consequences of any and all social actions would be predictable, which, of course, means that there would be no surprises and everything would go as planned. But the world is indeed imperfect, and we cannot foresee every reaction to every action. Every new invention or idea, whether it is in medicine or elsewhere in society, will have unintended consequences: unexpected things that go wrong. Therefore, the problem with body-head transplantation is not that there are latent outcomes; the problem is that there are so many. The aspects of BHT that have not been adequately addressed and that could go horribly wrong are so numerous that it is hard to imagine the experiment being conducted in any democratic country that strives to protect individuals from the hubris of institutions and unsafe practices generated by the imagination of social actors in the absence of social controls.

Pragmatic bioethics focuses on actions taken by individuals and institutions in a social and cultural context. Cultural definitions of right and wrong must be considered when evaluating emerging technologies. Doing so, however, means recognizing that standards of right and wrong are subjective and can vary by time and place. That China allowed body-head transplant research to continue where Western societies were reluctant to endorse the idea is a case in point. Ren and Canavero were confident that their ethical judgments on BHT are consistent with the broader moral schema, yet they realized that their judgments did not match the moral field of Western culture. So, they changed social contexts to one, China, that holds a different set of values, motivations, and institutional structures that see BHT as potentially normative and desirable.

Specifying morality in medicine is an exercise in a culture's sense of "what should be" and "what ought not be". Pragmatic bioethics is concerned with the moral consequences of collective actions (Shilling 2006), especially as they play out materially. We can speculate on the impact of BHT by looking at the potential beneficiaries of the surgery and the impact on the well-being of patients. In this context, pragmatic bioethics is more a method than a theory or ideology: it is a way of analyzing a problem. I have tried to stay close to the facts to see what results BHT might produce. Of course, facts are carved out of reality that is perceived and influenced by people's interests and motivations. In this light, it is no surprise to see that the advocates of BHT dismiss potential injurious psychological outcomes out of a pragmatic interpretation of the psychological literature that fits their intentions. It is the role of bioethicists to maintain a wholistic picture of the context in which a new idea is presented so that individuals and society are protected from what psychologists call motivated perception: seeing what we want to see and hearing what we want to hear.

A patient who is desperate to live and a surgeon who is inspired to explore unknown territory are tempted to adhere to a reality that conforms to those motivations. A reality that is skewed to mesh to desired ends causes individuals to overlook other "realities", facts that could derail their original intents and cause things to go wrong. Potential severe pain, tetraplegia, and memory loss are just three of those "other realities". Bioethics' duty is to protect individuals from shortsightedness that comes from tainted perceptions and articulate a response as to what ought not happen. This may be the case with BHT. Various steps of the process are currently only experimental. Psychological harm is not the only risk posed by BHT. The treatment of pain, fusing the spinal cord ends, and brain-body integration are unknowns at this point.

In Chapter 1, we visited Robert Merton's types of unintended consequences of social actions. Now that we have examined what body-head transplantation entails, we can return to Merton's ideas to help us see what can happen if BHT is allowed to occur. As we saw in Chapter 3, every Mertonian type of unintended outcome is likely if a BHT is attempted under the current proposal. The unintended outcomes constitute "what ought not happen" in a medical procedure and are sufficiently contrary to existing social norms of responsible medicine that it is in medicine's and society's interest to disallow the experiment. All but the most ardent advocates of conducting this experiment have expressed strong reservations about the efficacy of the BHT plan and its moral standing. Few who have written about BHT or with whom I have engaged in conversation about it are ambivalent or undecided about whether or not the surgery would work if it were attempted. Not many are unsure; most everyone has taken a position to accept or reject it. In my experience of reading and talking about this, the scales tip overwhelmingly to the latter.

One point of near universal agreement among critics is that the reconnection of a fully transected human spinal cord is not possible. Despite a few animal studies showing that PEGs have encouraged neuron growth, data purporting to prove that GEMINI works are not convincing. A complete neurological recovery in which a person can have normal control of gross and fine motor skills cannot be extracted from the animal studies. That a complete recovery has yet to be fully achieved in spinal cord injury repair adds weight to the conclusion that the technology to achieve this step of the BHT is unrealizable. The knowledge to heal any spinal cord injury is not available at this time in history. The critics are correct to assail BHT on these grounds. Leaving patients alive but paralyzed from the neck down and perhaps in terrible pain is cruel and medically and existentially in a worse state than before the operation.

Without the necessary research to show the credibility of the plan, the risk of committing errors is high. Relying upon untested and experimental interventions to help patients recover psychologically, for instance, is a mistake in light of all that is known psychologically and sociologically about the embodiment of the self. Ignoring over a century of scientific literature on the development of the self, as BHT advocates have done, is an intentional and somewhat inhumane response to a patient's likely emotional distress. BHT would be a physical and emotional trauma to a patient and one far too complex to rely upon an unproven novelty such as a virtual reality machine or an adjunct treatment like hypnosis to resolve. Imagine a scenario in which patients who are alienated from their own, but new bodies, and the best treatment they can receive is known at the onset to be experimental and insufficient.

Another error is the failure to consider how the brain will align itself with a new body. The lack of serious scholarship on the various mechanisms in which the brain and body will have to integrate indicates shortsightedness. The absence of this research is likely to lead to unexpected and negative results in the welfare of patients. The gut-biome system's relationship to mental health is but one example. BHT advocates themselves are uncertain about the impact of blending an old gut with a new brain, which is surprising given that we now have sufficient knowledge that the gut can influence mental health. It is likely that a new body alone would cause psychological distress to a BHT patient. Fecal transplants would not suffice to remedy this problem.

Another of Merton's sources of unintended consequences, imperious immediacy of interests, is particularly insightful to understand the unfolding BHT story. We have seen that BHT has taken on more meanings than its purported life-saving intent. In the discourse of the BHT "drama", the surgery has been linked to Chinese nationalism, colonizing Mars, immortality, and the futuristic notion of brain transplantation. That researchers wanted to conduct the surgery merely four years after introducing the idea in 2013 appears as a rush to conduct a surgery for which preparation had yet to be completed. In addition, acting out the narrative of BHT promotion on the internet and in the media and in the fashion in which it has been conducted, is suggestive of motives beyond altruism. There does seem to be a short-term game afoot, given the lack of research, the shallowness in responses to ethical criticisms, and the neglect of important aspects of the neuro-psychosocial aftermath. These are indeed uncomfortable considerations to discuss.

Sociologists are curious as to who benefits from particular social actions, and we can ask that about BHT. The accumulation of potential problems discussed in these pages means that the patient would not likely be the beneficiary in a BHT. Those who receive the attention and the money for conducting the research and executing the surgery are the true beneficiaries. Rather than do the research, there is a "let's do it and see what happens" feeling to this whole business. This may be good for those wanting to perform BHT but certainly bad for a patient. The initial rush to conduct the surgery felt like an instance of the old cliché that it is easier to ask for forgiveness than permission.

One of Merton's sources of unintended consequences also concerns basic values. In this case, the values underlying an action can produce outcomes that are different from or even in contrast to core values of both medicine and society. The value of autonomy in medical decision-making is one such example. Our belief in the right of individuals to decide their fates is so strong that we often neglect the social context in which decisions are really made. An unintended consequence of autonomy is disharmony between clinician and patient when the social context appears to impede what the clinician may believe to be the best course of action. In addition, it is easy to confuse autonomy with narcissism. There are times when individuals' desires are not realizable or healthy, despite their pleas for others to accept their decisions. We cannot consent to be murdered, and we cannot consent to undergo a medical procedure that could cause us great harm if the medical and bioethical community believes its costs outweigh any possible benefits. Such is the case here. The medical establishment is not anti-innovation, but medicine does stand for rational responsibility in the innovation, practice, and the protection of patients.

Of course, saving human life is the most cherished value in medicine, but at what cost? To put it bluntly, a consequence of a body-head transplant is the death of up to eight other terminally ill patients who could benefit from receiving the organs of a single donor; all of whose organs went to save the one who received the BHT. Such a practice is a high price to pay to save one life and speaks to elitism. One patient might benefit from the BHT, but eight will pay for it with their lives. This is not an economy of responsible or ethical practice.

Merton termed his last source of latent functions self-defeating predictions. This concept describes the case when public prediction of a social development proves false because the prediction itself changed the course of events. The unintended consequence is that the fear of the consequences of a social action propels efforts to find solutions before the problem occurs. It is possible that a rise in the number of proposals for new yet bizarre and dangerous yet potentially lifesaving surgeries such as BHT will lead to greater control of new surgical interventions akin to the FDA's oversight of new drugs. While this might feel anathema to surgeons, such an action should not impede everyday surgery and slight alterations in established procedures. What could happen is that new unproven surgeries that push the boundaries of extant moral standards, as BHT does, would require FDA-like approval. If surgeries such as BHT continue to be proposed, society may demand greater regulation of those procedures.

The absence of empirical support for many of the steps in BHT and the reliance on unproven experimental techniques provide numerous outcomes beyond the one that is desired: the restoration of complete health for the head donor. What makes them unintended is that they have not been addressed although we have the ability to know that the potential for complications is high. Here I am primarily speaking to brain-body integration and emotional and cognitive distress that verge on the traumatic.

Are Body-Head Transplants the Same as Face Transplants?

In making their argument that BHT will not exert any undue psychological stress on patients, Ren and Canavero (2017a) likened body recipients to face transplantees, who, as studies show, have emotionally and socially benefited from their new faces. Indeed, although we discussed this in Chapter 1, there are additional cautions that should be raised here. First, not every face recipient's psychological well-being has been conveyed via scholarly publications, which means we cannot say that every face transplantee is doing well emotionally. Perhaps they are, but we cannot say that without knowing more about the cases absent from the literature.

Second, facial disfigurement is not life-threatening; hence, face transplantation is life-enhancing. Where we see improvement in psychological health among face recipients it is their perceptions of their improved appearance that reduces the anxiety they feel in social encounters and when looking at themselves. Body image is a powerful emotional component of one's sense of self, and when a person's face is dramatically distanced from the social norm, as disfigured peoples' are, depression, anxiety, and low self-esteem are common. This situation is not parallel among people who would want a body-head transplant.

Third, receiving a face, which includes skin and its underlying tissues, is not equivalent to receiving an entire body either in scale or body tissues transplanted. Therefore, the psychological profiles of face recipients are not generalizable to hypothetical body-head transplantees. Although most face recipients may be doing well emotionally, the physical aftermath has not been as encouraging. As shown in Chapter 1, sadly, roughly one in six face recipients has died, including one suicide. If adjusting to scale—face to body—face transplantation is not a encouraging predictor of the odds of surviving a body-head transplant.

For these reasons, deducing that the psychological well-being of BHT patients would be improved similarly to face recipients can be dismissed.

But this comparison of face transplants to head transplants raises an interesting ethical question. In the early days of face transplantation research, critics of the concept were admonished for being patriarchal, for telling patients that receiving a new face was not good for them. These critics contended that facially disfigured people should not carry the burden of society's rejection of them by undergoing an unproved surgery and then taking high doses of immunosuppressive drugs. Research, as discussed in Chapter 1, countered the detractors' editorial position by finding that people with facial disfigurement, as well as kidney transplantees who were already on these drugs, were willing to take them in exchange for the benefits that the transplants offered. Are we doing the same thing with body-head transplantation? Are we telling people what is in their best interest by saying that BHTs should not be done?

As the case has been made that face and head transplants are not equivalent, the question essentially amounts to little more than a *non sequitur*; the two surgeries are in no way parallel procedures. For one, the stakes are far higher with BHT because the patients who would undergo the surgery are dying, whereas face recipients are not. What's more, the risks are vastly different between the two. Plus, the head recipient is receiving an entire body filled with organs, bones, muscles, DNA, gut bacteria, and a host of other biologics. Facial appearance, of course, will not change for BHT patients, but their body composition will, and body image is as equally important to identity as facial features are to an individual's psychological profile.

The surgeries are different as well. Whereas face transplantation, up to a certain point, has an exit strategy, BHT does not. The technology for face transplant surgeries largely required relatively little invention of new techniques, other than immunosuppression. Established techniques were applied to the face. While the head transplantation plan also includes many established techniques, its most significant steps are comprised almost completely of untried procedures such as decapitation, head reattachment, and spinal cord fusion. Face transplantees do not have to worry about organ failure; BHT patients have a risk of *every* organ in the new body failing or being rejected. Rejection of tissues in the face are visible in the form of a rash, but organ rejection is detected through illness. Multiple organ or tissue rejection would be a nightmare and quite possibly fatal for BHT patients. The list of dissimilarities goes on and on. Body-head transplantations are far more dangerous, and, as we have seen, the research necessary to comfort patients and the medical community that the risks are clinically manageable does not exist at this time. In a nutshell, the volume of potential error produces too many opportunities to fail.

If we add to the mix that a successful BHT does not guarantee psychological health, and in fact it portends psychological troubles, we can safely conclude that the medical community does have the responsibility to tell patients that this procedure is not in their best interest. If some critics are even questioning the legal standing of the surgery, there are clearly many obstacles yet to overcome before BHT is allowed.

Patients do not always understand everything their physicians tell them. Though we gift our doctors with both the responsibility and the expectation that they can cure us of every ailment and rescue us from death, many patients come to believe that anything is possible. Is it our narcissism that makes us feel entitled to a long, pain free life? Do we really believe that anything is possible based on the say-so of some person of supposed authority? Our fear of death can blind us to accept offers that may sound wonderful and miraculous but have no such guarantee. It is the responsibility of the bioethical trust within medicine and general society to monitor what is real and that which is safe. Just as we rely on experts to cure us, we must also rely upon experts to monitor those who say they can cure us.

Mind and Body Integration

The lack of attention BHT researchers have paid to psychological well-being borders on negligent. Referring to the self as an illusion is a reckless stance that ignores what makes patients human.

Agreed, the self is not a material thing. It has no corporeal presence nor material form. The self is not part of the brain or body, it *is* the brain and body working as an indissociable unity integrated by biochemicals and neural circuits functioning systematically and in unison (Damásio 1994). Our bodies are experienced as intimate parts of our lifeworld, our humanness (Cherry 2022). To speak of myself is not to speak of a self that I have, but simply to speak of the human being that *I am* (Bennett and Hacker 2003).

We can lose a body part and gain a new one without significant loss to our sense of self. Replacing a heart, foot, kidney, or face with a natural or artificial "part" does not destroy our sense of embodiment and identity as a person. Still, we have to accommodate the change, which can alter our identity however slightly. Many organ recipients think of themselves as "us" rather than just "me", and their activities are often altered because of the post-operative limitations placed on patients. If we destroy too much of our bodies, restrictions on activities that were important to the patient's identity may impact their psychological recovery. Athletic individuals with various prostheses are usually told or forced to limit certain sports or movements and organ recipients on immunosuppression must be hypervigilant to avoid contact with contagions, which could limit their social lives. Most cope and adapt to their new life conditions and their sense of personhood remains intact, and society has created new opportunities for people to continue to prosper and find self-fulfillment.

Mark Cherry (2022), however, contends that BHT patients could have more trouble incorporating their new bodies into their self-image. Phenomenologically, accepting an entire body may prove too hard to maintain over time, and the physical separation between the mind and the body may prove too significant to overcome. The ability to answer the question "who am I?" will not simply work itself out; it may become increasing arduous as the foreignness of the body becomes more apparent if integration does not occur smoothly. The disruption of the personal narrative of the patient is more likely to be traumatic than readily solved. A natural, smooth transition of the embodied self as perceived by the mind is not a given.

The body contributes to the development of the self through its biological role in influencing the brain and the socially infused meanings of the body. The health and strengths of vital organs and the body overall affect self-identity. People, especially children, are stigmatized for wearing glasses; short people lose jobs to tall people, all things being equal, because of the perceived attributes of height; seniors are often reluctant to get hearing aids for fear of losing their youthful appearance and status; and aging itself so alienates some individuals from their bodies that they undergo expensive cosmetic surgeries in attempts to create a new appearance, one that is younger looking and hopefully more attractive in the eyes of society.

Furthermore, our bodies are highly individualized (Shiller 2013). We style them to fit our perceptions of ourselves as we are or who we want society to think we are, and sometimes in reaction to social pressures. We get tattoos, brandings, piercings to create an image, externalizing our inner desires and self-image, and to conform to the expectations of a particular social reference group, all while calling these markings "self-expression". Prejudices against certain physical characteristics lead people to use chemicals to lighten their color of their skin or straighten naturally curly hair. We exercise to keep healthy, but also to enhance the perception of our sexual desirability to others, and the money spent on clothes and jewelry to adorn our bodies to enhance or hide our real social status has made fashion a major world industry.

Pressures are mounting in contemporary society to end so-called "body shaming", the act of humiliating and embarrassing someone because their body does not conform to socially constructed ideals of beauty, purity, or correctness. The fact that so many people remain disillusioned with their physical presence or bully others because of theirs tells us that we have much work to do to eliminate the stigma of physical differences. For people who do not conform, feeling shame or out of favor with standards of the social hierarchy are common if not every day experiences.

Cognitive sciences have suggested that human cognition does not solely originate within the brain; rather, humans exhibit an embodied cognition in which our bodies participate in the formation of self (Cuoco and Davey 2016). Psychologists have described embodied cognition as cognition that is dependent on the various kinds of sensory experiences that a person's body has. Further, they note that sensorimotor capabilities are central to a more comprehensive biological, psychological, as well as cultural context (Varela et al. 1991). Consequently, difficulty in adjusting to new sensory stimulations may prove psychologically difficult and impair adjustment to a new body. Joshua Cuoco and John Davey are also concerned that the individual's personality and memories may be dramatically altered after a body-head transplant. They conclude that:

> the procedure of human head transplantation may potentially serve as an exchange of a debilitating disease for psychological confusion and instability and serious psychological complications such as mood disorders, suicidal thoughts, and psychosis (Cuoco and Davey 2016: 2).

One step that is necessary before BHT could ever go forward is that advocates must formulate a theoretical and clinical response to these potentialities rather than dismiss them or propose gimmicks and unproven therapies. The problem they face is that such a treatment plan would be long, complicated, and also unproven.

Habits, Muscle Memory, and the Social Construction of the Body

People have habits, conscious and unconscious behaviors that are nestled in the brain but acted out by the body. These habits are so repetitive that the brain and body can execute the behavior without the "mind" thinking about doing them. Although muscle memory is seated in the brain, a muscular-neural connection is built through repetition that would be lost when the original body of the head donor is removed. Habitual behaviors are crucial for daily living and earning a living, and losing these recurring motions would not be trivial matters for BHT patients. They would have to re-teach their new bodies not only how to perform these behaviors but also that they must done.

Habits serve a number of different purposes, some of utilitarian value and others that are of more emotional necessity. All require muscle memory to various degrees, and all would have to be relearned.

Utilitarian habits such as the motions in brushing one's teeth, typing, or riding a bicycle are learned so well that we do not have to think about every action or motion—they come about somewhat automatically. Many habitual behaviors and their muscle movements are complex. People at their jobs are often on "auto pilot", and athletes, artists, and musicians are so well trained that they can function at a high level of performance on muscle memory alone, allowing them to think about creativity and expression rather than how to spin clay on a potter's wheel or recall the difference between a *pirouette en dedans* and a *pirouette en dehors*. Creative people, including surgeons, get into "the zone" where their focus is on interpretation, inspiration, and performance, not basic movements. Musicians are not remembering how to finger the notes on their instrument; they are centered on expressing that note in the context of the entire piece of music they are playing.

Other habits are more subtle and are loaded with emotional value rather than occupational or artistic action. Many of these habits are mindless stress-reactive behaviors that we may know or not know we do. Clenched shoulder muscles or hands indicate unconscious, or conscious, anxiety. Using drugs or smoking tobacco are habits that often begin as self-medicating responses to emotional problems or stress.

Routines and habits, as Chris Shilling (2008) says, are necessary behaviors that can enlarge or restrict our relationship to the world. They connect us to a social environment loaded with near constant behavioral expectations. We cannot mentally process all of those expectations. There are too many things we are expected to do in today's world.

What will happen if a BHT patient could not handle the new body's inability to meet most or even all of those expectations? This problem would not be mere irritation. Frustration would certainly arise, and if the disconnect between the patient's expectations and abilities were prolonged, depression, anxiety, body alienation, and dysmorphic reactions could easily follow. The dissonance between the demands of the social surroundings and the desires of the mind may be too stressful to overcome, making experimental gimmickry and hypnosis insufficiently potent tools. Instead, a deeper, more sophisticated psychological treatment, more akin to what is used to treat severe trauma, would be necessary.

At the heart of the matter is a loss of the command of the body, a body that has no social, psychological, or medical history with the mind and personal identity. Losing those taken-for-granted habits of the body and then having to relearn them after a long period of fear of dying, recovery, and pain would be further traumatic and likely initiate a crisis of identity. The loss of the embodied self is an existential disruption of the highest order and will disturb the very core of how individuals interact with the world around them.

None of us has a perfect body. We each have a flaw of some sort that separates us from the cultural ideal. For many people, their imperfections, relative to that ideal, are nominal or perhaps imagined. For others, their corporeal deviance is more noticeable and subject to stigma. It is not uncommon for individuals to try to conceal their physical irregularities or non-conforming characteristics, whether trivial or conspicuous, and try to "pass" as normal.

The subjective aspects of the body are quite real and highly meaningful, often representing fundamental beliefs and group memberships that define ourselves and give order to the world. Would an observant Jew or Moslem take a body long nourished by eating pork? A woman who is against abortions might find accepting and living with a body that had terminated a pregnancy emotionally straining, creating a conflict of conscience. Accepting the perceived flaws in the new body could be harder than many think. The new one will not be perfect and may represent someone else's real or imagined inadequacies and faults. We cannot answer these questions with the extant BHT literature.

We rely on our bodies to engage our surroundings and to express our inner thoughts and desires. In that sense, the body is a social construct (Adelman and Ruggi 2015). Our bodies engage the social environment through labor, recreation, and relationships and are perceived as the physical structure of our inner selves. Through this engagement, our bodies also impact our psychological character. We work to make our bodies into the image of our social roles and statuses while simultaneously our bodies are created and defined by the social order. Race, for example, is a social construction. In sociology we often say that race is a biological concept with a sociological definition. Race has little significance outside its social meaning. As real "things", sex and age are also steeped in social constructions to give them meanings well beyond their biological denotations. Because of all the sociological interpretations of the body physical, many people feel self-conscious about their bodies if they do not match social standards, and in extreme cases they may engage in self-destructive thoughts and behaviors if their bodies are "different".

In sum, our bodies are important to who we are as subjectively thinking human beings. They are the physical presence and representation of our humanness and critical components of our examinations of ourselves. Our bodies are the material form of self-referencing, the process in which we recognize how our own thoughts and actions are intertwined with our surroundings. The body is more than a visible container, it is simultaneously an expression and shaper of the mind. How patients, regardless of their performance in pre-operative psychological evaluations, would cope with losing that part of their humanness and at the same time incorporate a new body, that of another's intact self, seems daunting and certainly more difficult than some would think.

The message here is that the body is vitally important to the formation of self-image and self-esteem. People's embodied self, their appearance and health, affect their identity and self-efficacy, and it is naïve to dismiss the corporeal dimensions of our self-perceptions.

It is highly possible that the mind will not accept the new body as "me", regardless of whether the brain does.

Studies have reported that some organ transplant patients have stated that they have retained some of the tastes and preferences of their organ donors. A few have reported their tastes in food and music and sexual orientation were different after their transplants and aligned with their donors. Others claim to have assumed personality and artistic characteristics of their donors (Liester 2020). Despite these accounts, physicians and researchers typically react to these testimonies with skepticism since there is no known direct causal mechanism to transfer memories believed stored outside the brain. Nonetheless, such tales are not uncommon. Could it be that cellular memory effects personality changes in organ transplantees? Michael Liester (2020) has proposed that hypothesis. Memories of the donor's life, he writes, could be stored in the cells of a donated heart and then recalled after transplantation. He proposes that these memories could be saved as cellular memory — epigenetic memory, DNA memory, RNA memory, and protein memory. It is an intriguing proposal, one that has been rudimentarily demonstrated in DNA experiments with worms (Benayoun and Brunet 2012).

If there is something to cellular memory, the implications for BHT are not small. An entire body of "memories" and DNA activity that may or may not mesh with the new brain could have physical and psychological implications, though not necessarily for the betterment of the body recipient. From Liester's review of studies on organ recipients taking on characteristics of their donors, we also learn of accounts of psychopathology being transferred from donor to recipient. That finding may be consistent with gut-biome dissonance just as easily as cellular memory effects. We do not have an answer to these questions at this time, but we do have the question. And it is a question that has not been addressed and could result in a serious unexpected outcome of a body-head transplant surgery.

Who Am I?

The discussion about the nature of the embodied self speaks directly to the worries expressed by many critics that the BHT patient will not be a continuation of the self that exists in the brain but will be a "new person". The self that is the mind is rooted in the self that is the physical, and there is no reason to believe that there will not be a disruption in the self as it was before the transplant. Whether or not one is Confucian or Buddhist or steeped in the Aristotelian logic of the Western tradition, the mind is entwined with the body's well-being and its appearance. Regardless of the philosophical or epistemological understandings of the connections between mind and body, they are entangled in every aspect of life: biomedically, psychologically, and sociologically. The BHT survivor would truly be a new person in mind, spirit, and body.

I am curious as to how we would become a new person and reconcile the loss of the "old me". The loss of their body would require BHT patients to grieve over their own partial death. Saying goodbye to one's body and perhaps even attending its funeral would require a presence of mind that few people possess. As I sit here thinking of this and trying to imagine what it would be like to bury or cremate my own remains, I feel an intense sadness bordering on panic. (I am not ordinarily prone to such immediate turns of emotions, by the way). A sense of fear has also come over me as I visualize my body in the ground, headless and unrecognizable to my family and friends who know what lies in the shroud I have chosen for my own final rest. But I am not at rest; part of me remains alive. I already feel alienated, and all I am doing is sitting in a coffee shop engaging in fantasy.

People who survive a BHT will have to imagine their new bodies as their own. They will have to reconcile that their old physical presence is dead and buried and their new body is now "themselves". They will have to decide who they are, and the process by which they come to any acceptable conclusion will be the most difficult psychological adjustment ever attempted. The question, "Who Am I?", will be confusing because there will be no reference point for self-identity. Self-concept is largely based on the body's appearance, attributes, and social placement. How BHT patients will detach themselves from the self-concept grounded by their old bodies in order to form a new one based on a new body and status will likely cause cognitive dissonance and emotional distress. Patients, I suspect, will not easily know who they are after undergoing body-head transplantation.

All in all, after thinking long about BHT, I remain unconvinced in Čartolovni and Spagnolo's prediction that BHT would lead to certain madness. Nonetheless, it is possible, and given all the adjustments that the mind and brain would be asked to make, BHT patients would at least experience a traumatic crisis unlike any ever recorded. This trauma would stem from an impaled identity, one pierced by the intrusion of a new physical presence acting like a spear jabbing at the mind. I predict discombobulation and perhaps a state as close to the complete disintegration of the mind as can be. The struggle to reconcile "Who I am" with "what I am" could push patients to the edge of madness. The treatments proposed to ease that distress, however, would be so inadequate that they could push patients over the edge and possibly prove Čartolovni and Spagnolo correct and cause patients to go utterly insane.

Last Words

The ethical complications of BHT are profound; they are a spaghetti bowl of discontinuity, misdirection, and shortsightedness. As far as I can see, decapitating the head of a desperately ill individual and attaching it to a healthy body has no ethical solvency or currency. Most expectations of normative medical and psychological practice would be violated if a BHT were conducted, and many, such as humane animal research and open transparent communication about risks, already have. BHT is corrupt in theory and practice and has little moral fecundity. Yes, BHT is potentially life-saving, but to perform one BHT would abandon as many as eight other good souls to their death because the organs they need have selfishly been assigned to a single person. Organ donation follows a priority system in which recipients are ranked on the severity of their illness, time on the waiting list, and geographic location. Patients who could benefit from a body-head transplant, if the procedure ever became a viable and safe option, should be low on that hierarchy of recipients, rising to the top only when the need for the body donor's organs no longer exists. A future in which organ donation is no longer needed to save lives is one that we cannot yet imagine.

Does criticizing the ethical soundness of BHT really matter? It is easy to fall back on the Polish expression: *Nie mój cyrk. Nie moje małpy.* "Not my circus. Not my monkeys". I am not interested in getting a new body, so why should I be dragged into this quagmire of risks and unusual promotion of a medical innovation. What do I care if someone wants to undergo a body-head transplant? Well, it is not so simple. We all have a stake in whether BHT is found morally acceptable. BHT is a test of our collective moral presence and recognition that modern science or any other knowledge system has its boundaries. Cheating death is a short game. The long game is to build a code of morality that resists narcissistic greed and favors the common good. It is the game we must win.

References

Adelman, M. and L. Ruggi. 2016. The sociology of the body. Current Sociology, 64(6): 907–930. doi.org/10.1177/0011392115596561

Benayoun, B.A. and A. Brunet. 2012. Epigenetic memory of longevity in Caenorhabditis elegans. Worm, 1(1): 77–81. doi.org/10.4161/worm.19157

Bennett, M.R. and P.M.S. Hacker. 2003. Philosophical Foundations of Neuroscience. Blackwell Publishing, Hoboken, NJ.

Cherry, M.J. 2022. What happens if the brain goes elsewhere? Reflections on head transplantation and personal embodiment. The Journal of Medicine and Philosophy: A Forum for Bioethics and Philosophy of Medicine, 47(2): 240–256. doi.org/10.1093/jmp/jhab045

Cuoco, J.A. and J.R. Davey. 2016. Operation Frankenstein: Ethical reflections of human head transplantation. Insights in Neurology, 1(2): 1–4.

Damásio, A.R. 1994. Descartes' Error: Emotion, Reason, and the Human Brain. Avon Books, New York.

Liester, M.B. 2020. Personality changes following heart transplantation: The role of cellular memory. Medical Hypotheses, 135. doi.org/10.1016/j.mehy.2019.109468

Ren, X. and S. Canavero. 2017a. HEAVEN in the making: Between the rock (the academe) and a hard case (a head transplant). AJOB Neuroscience, 8(4): 200–205. doi.org/10.1080/21507740.2017.1392372

Shilling, C. 2006. The Body and Social Theory, 2nd ed. Sage, London.

Shilling, C. 2008. Changing Bodies: Habit, Crisis and Creativity. Sage, London.

Varela, F.J., E. Thompson and E. Rosch. 1991. The Embodied Mind: Cognitive Science and Human Experience. MIT Press, Cambridge.

Index

For Product Safety Concerns and Information please contact our EU
representative GPSR@taylorandfrancis.com
Taylor & Francis Verlag GmbH, Kaufingerstraße 24, 80331 München, Germany